A SUBJECT WITH NO OBJECT

D0713999

A Subject with No Object

Strategies for Nominalistic Interpretation of Mathematics

JOHN P. BURGESS

and

GIDEON ROSEN

CLARENDON PRESS · OXFORD

OXFORD

UNIVERSITY PRESS

Great Clarendon Street, Oxford OX2 6DP

Oxford University Press is a department of the University of Oxford.
It furthers the University's objective of excellence in research, scholarship,
and education by publishing worldwide in

Oxford New York

Athens Auckland Bangkok Bogotá Buenos Aires Calcutta
Cape Town Chennai Dar es Salaam Delhi Florence Hong Kong Istanbul
Karachi Kuala Lumpur Madrid Melbourne Mexico City Mumbai
Nairobi Paris São Paulo Singapore Taipei Tokyo Toronto Warsaw

with associated companies in Berlin Ibadan

Oxford is a registered trade mark of Oxford University Press
in the UK and in certain other countries

Published in the United States
by Oxford University Press Inc., New York

© John P. Burgess and Gideon Rosen, 1997

The moral rights of the authors have been asserted

Database right Oxford University Press (maker)

First published 1997
First issued as paperback 1999

British Library Cataloguing in Publication Data

Data available

Library of Congress Cataloging in Publication Data
Burgess, John P.
A subject with no object : strategies for nominalistic
interpretation of mathematics / John P. Burgess and Gideon Rosen.
Includes bibliographical references and index.
1. Mathematics—Philosophy.   I. Rosen, Gideon A.   II. Title.
QA8.4.B86 1996 510'.1—dc20   96-28175

ISBN 0–19–823615–8
ISBN 0–19–825012–6 (Pbk.)

Printed in Great Britain
on acid-free paper by
Bookcraft Ltd.
Midsomer Norton, Somerset

PREFACE

'A preface is not, in my book, an introduction.' So a prominent philosopher of mathematics once wrote. To judge from his practice, he took the function of an introduction to be to begin the exposition of his subject, and that of a preface to place it in its intellectual context. We will follow his example.

For all its wealth of results, and for all the power of its applications, mathematics as of about 1800 dealt with only a handful of mathematical structures, all closely connected with the models of time and space used in classical physics: the natural, rational, real, and complex number systems; the Euclidean spaces of dimensions one, two, and three. Indeed, mathematics was widely held to deal directly with the structure of physical space and time, and to provide an example of pure thought arriving at substantive information about the natural world. The central question in philosophy of mathematics at that period was how this could be possible. All that changed completely during the nineteenth century with the introduction of more and more novel mathematical structures, beginning with the first non-Euclidean spaces.

Among other consequences, the proliferation of novel structures that then ensued made it seem desirable to insist on a closer adherence than had become customary to the ancient ideal of rigour, according to which all new results in mathematics are to be logically deduced from previous results, and ultimately from a list of explicit axioms. For in the absence of rigour, intuitions derived from familiarity with more traditional mathematical structures might easily be unconsciously transferred to novel structures where they are no longer appropriate.

The introduction of so many new structures naturally tended to lead to increased specialization among mathematicians. However, the tendency to fragmentation was powerfully counteracted by the emergence of new cross-connections among its various branches, arithmetic, algebra, analysis, and geometry: a broadening of the notion of algebra, for instance, allowed mathematicians to recognize algebraic structures connected with geometric objects; a broadening of the notion of geometry allowed them to recognize geometric structures connected with analytic objects; and so on. Because of these cross-connections, rigorous axiomatic treatments of the several branches of mathematics separately would not suffice: a unified, general, rigorous framework for all the interconnected branches of mathematics was wanted. The search for such a framework led to intensive activity in philosophy of mathematics by the early decades of the twentieth century.

By that time there had emerged, from work on rather specialized questions raised by the generalizing tendency in analysis, a theory of sets of real numbers or linear points, and finally a theory of general sets of arbitrary elements. When controversy and confusion—indeed, contradictions—arose in that theory, it was reformed and reformulated on a rigorous, axiomatic basis. Axiomatic set theory provides a unified, general framework for mathematics, and one conforming to the ideal of rigour. It does not, however, conform to the ideal that axioms should be self-evident truths. It was only when it came to be widely perceived that various mathematico-philosophical programmes promising self-evidence and certitude had not succeeded, and were unlikely to succeed, that mathematicians acquiesced in axiomatic set theory as a framework for their subject. Philosophy of mathematics then entered a period of comparative quiescence that lasted through the middle decades of the century.

In these closing decades there has been renewed activity in philosophy of mathematics. Many recent and contemporary philosophies of mathematics challenge the accepted format for mathematical work: rigorous deduction of new results, ultimately from set-theoretic axioms. Often the challenges are revivals of positions taken by some during the previous period of philosophical activity around the turn of the century, updated to take account of subsequent developments in philosophy, mathematics, logic, and other disciplines. Thus some propose to expand the accepted framework by adding new axioms to set theory, to be justified either intrinsically as reflections of some intuitive conception of set, or extrinsically on account of their attractive consequences. Others propose to move beyond the accepted framework in a quite different direction, the relaxation of the standards of rigour, so as to admit in the context of justification in pure mathematics what has always gone on in the context of discovery and in applications, a combination of heuristic argumentation with inductive verification of special cases, by hand or by machine. In contrast to these two divergent expansive proposals, there has been a revival of the restrictive proposals of constructivism, a loose cluster of schools of thought whose least common denominator is that they reject most of set theory, and insist on tighter standards of proof than do the orthodox: they reject non-constructive existence proofs, proofs that purport to establish the existence of a mathematical object with a certain mathematical property, but that do not provide any specific example.

In contrast to all the positions mentioned so far stands a quite different restrictive philosophy, nominalism. Nominalism (as understood in contemporary philosophy of mathematics) arose towards the mid-century when philosophy of mathematics was otherwise rather quiet. It arose not in the mathematical community, as a response to developments within mathematics itself, but rather among philosophers, and to this day is motivated largely by the difficulty of fitting orthodox mathematics into a general philosophical account of the nature of knowledge. The difficulty largely arises from the fact that the special, 'abstract' objects apparently assumed to exist by orthodox mathematics—numbers, functions, sets, and their ilk—are so very different from ordinary, 'concrete' objects. Nominalism denies the existence of any such abstract objects. That is its negative side. Its positive side is a programme for reinterpreting accepted mathematics so as to purge it of even the appearance of reference to numbers, functions, sets, or the like—so as to preserve the subject while banishing its objects. Actually, there is not one programme of nominalistic reinterpretation or reconstrual, but several, since nominalism is a loose cluster of positions, and different nominalists prefer quite different strategies and methods.

The literature on nominalism is rather large, partly because of the diversity just mentioned of strategies proposed, partly because the execution of any one strategy to the point where one has reasonable assurance that it can be made to work tends to require a book-length work, or a series of articles of comparable bulk. The result has been the creation of a substantial shelf of works on nominalistic reconstrual. Considered in relative rather than absolute terms, this body of work forms an especially large part of the recent and contemporary literature on philosophy of mathematics. For the field of philosophy of mathematics today tends to be regarded among professional philosophers as a legitimate if not especially prestigious area of specialization, and among professional mathematicians as a dubious though probably harmless hobby. Naturally, then, writings by philosophers about their concerns tend to outnumber writings by mathematicians about theirs in the recent and contemporary literature of the field; and nominalism is distinctively a philosopher's concern.

Being the work of philosophers rather than logicians or mathematicians, the various nominalistic projects are characterized more by philosophical subtlety than by technical virtuosity. The difficulty for students and other interested readers does not lie (or should not lie) in technical aspects of the individual works, which should seldom require a knowledge of more than freshman-level mathematics and introductory-level

logic. Rather, the difficulty lies in the fact that the works are widely
scattered in various books and journals, are written in different notations
and terminologies, and are each devoted to the development of some one
preferred nominalistic strategy, with little time or space being spared for
comparison with rival approaches. As a result of these and other factors,
there has been some unnecessary confusion and controversy over technical matters, and preoccupation with such matters has tended to distract
attention from more properly philosophical concerns.

The primary aim of the present book is to provide a comparative survey of the various nominalistic projects, clearing up technical issues,
in so far as this is possible in the present state of knowledge, and identifying without pretending to settle philosophical issues calling for further
reflection. The long Introduction, Chapter I.A, discusses the philosophical background. What considerations motivate philosophers to
endorse nominalism? What considerations motivate nominalists to undertake reconstructive projects? The central chapters of the book undertake
a comparative study of such projects. Chapter I.B introduces a general
formal framework. The three chapters of Part II present strategies deriving
from the work of three contemporary philosophers of mathematics, Hartry
Field, Charles Chihara, and Geoffrey Hellman. Chapter III.A provides
a potpourri of further strategies. Chapter III.B provides a partial guide
to the literature. The short Conclusion, Chapter III.C, attempts a provisional assessment. We believe, for reasons partly articulated there, that
the strategies of nominalistic reconstrual should be of interest even to
anti-nominalists. (As will become abundantly clear before the reader is
very far into the book, we are neither adherents of nor sympathizers with
nominalism ourselves.)

A prominent philosopher of mathematics has written that 'philosophical
hostility to abstract objects . . . springs from a kind of superstition and
. . . no good purpose is served by ingenious attempts to purge our language
of apparent reference to any but concrete objects.' We believe, however,
that even one who accepts the first half of this assertion should not accept
the second.

ACKNOWLEDGEMENTS

In the course of over a dozen years of thinking, speaking, and writing about nominalism, I have had the benefit of important comments from quite a number of persons, whom I have tried to thank in my successive publications during the period. I would like to extend to them collectively my renewed thanks, even though I will not try here to list them by name, for fear of inadvertent omissions. What I must do here is to thank individually some of those who have been a constant source of help throughout the whole period in question.

These include, first of all, the three philosophers mentioned in the Preface, who have provided several forms of assistance, beginning with their making available their works in preprint prior to publication. With Professors Chihara and Hellman I have moreover had many enlightening philosophical discussions over the years; in the case of Charles Chihara, these go back to my days as his student at Berkeley.

I have also greatly benefited from discussions and encouragement over the years from such colleagues as Gil Harman, Dick Jeffrey, and Bas van Fraassen. My colleagues Paul Benacerraf, David Lewis, and Saul Kripke have especially influenced my thinking about matters treated in this book, to a greater degree than will be apparent from the occasional mention of their names and works in the body of the text. Doubtless the book would have been better had I been even more receptive to their influence. Likewise I have had much benefit from philosophical discussions with my students, Peter Hanks and Mark Kalderon, though the former's senior thesis, 'Against the Abstract/Concrete Distinction', and the latter's doctoral dissertation, 'Structure and Number', were completed too late for me to make full use of them in the present work.

I owe a very special debt to Penelope Maddy. Some years ago we undertook to write jointly a work surveying both the more technical and the more purely philosophical aspects of both nominalism and the various schools of constructivism. When we had each finished a first draft of a fraction of the projected book, we realized that to complete the plan on the scale we thought appropriate to the subject would involve writing not a book, but an encyclopedia, and therefore set aside our original plan for a comprehensive joint work in favour of separate publications. My draft survey of the technical side of nominalism provided the nucleus for

Chapters I.B to III.B of the present book. I benefited greatly from Maddy's advice and encouragement while preparing that draft, though she is not responsible for the final form in which the material appears here. She did read the completed manuscript, and provided a useful list of corrections.

On a more person level, I am grateful to my wife, Aigli Papantonopoulou, and my sons, Alexi and Fokion, for their patience with an often somewhat distracted husband and father during the period of the writing of this book. My contribution to the book should be regarded as dedicated to my parents, Violet Patton Burgess and Arthur James Burgess.

J.P.B.

If intellectual debts are abstract entities—and what else could they be?—then the fact that I have so many of them ought to serve by itself as an altogether convincing refutation of nominalism. My contribution to the present volume, almost entirely confined to the non-technical Chapters I.A and III.C, derives from my doctoral dissertation. My chief debt is therefore to my supervisor, Paul Benacerraf, who first introduced me to the problems of nominalism and for whose guidance and encouragement I am deeply grateful. I owe a similar debt to my teachers (now colleagues) at Princeton, in particular David Lewis, Bas van Fraassen, and Mark Johnston, whose very different approaches to metaphysics and epistemology have each in one way or another shaped my own approach to these subjects.

A great number of friends and colleagues have helped me with this material: Steve Yablo, Sally Haslanger, Jim Joyce, Mark Kalderon, Jim Lewis, Peter Railton, Jamie Tappenden, Paul Boghossian, Nick White, and Larry Sklar. I am particularly grateful to Hartry Field and Crispin Wright, both for their writings on these subjects, which have provided a constant and welcome challenge, and for their generosity in discussing these matters with me.

My contribution to this book is dedicated to my wife, Lisa Eckstrom, for whose love and friendship I could not be more grateful.

G.R.

CONTENTS

PART I

Philosophical and Technical Background

A

Introduction

a. . . . of This Book

Finally, after years of waiting, it is your turn to put a question to the Oracle of Philosophy. So you humbly approach and ask the question that has been consuming you for as long as you can remember: 'Tell me, O Oracle, what there is. What sorts of things exist?'

To this the Oracle responds: 'What? You want the whole list? Look, I haven't got all day. But I will tell you this: everything there is is concrete; nothing there is is abstract. Now go away and don't bother me.'

This is disappointing: you had hoped to be told more. Still, you feel you ought to be grateful to the Oracle for telling you as much as she did, and you are firmly convinced of the truth of what you have been told. So you come away a believer in concrete entities alone; you come away a disbeliever in abstract entities. You come away a **nominalist**, in the most common contemporary sense of the term.

And yet, as you begin to reflect on the implications of nominalism for your intellectual life, you may come to be thankful that the Oracle didn't tell you any more than she did about what there isn't. For you are now constrained by a firm conviction that you ought to resist any account of the world, whether its source is philosophy or science or common sense, that involves abstract or non-concrete entities, in the sense of asserting or implying or presupposing the existence of entities of a sort that philosophers would classify as 'abstract' rather than 'concrete'.

In some cases, of course, this resistance will be easy. Presented with Plato's theory of Forms or Archetypes, the very Form and Archetype of a theory involving nominalistically objectionable entities, it will be no great strain to say now, as you may well have been inclined to say before: 'I reject Plato's metaphysics in the same sense in which I reject Aristotle's physics. I neither put any belief in it nor make any use of it myself, nor will I condone others doing so.' In other cases, however, the problem will be considerably more delicate. For everyday talk—to say nothing of technical science—is littered with assertions apparently involving nominalistically objectionable entities of the most diverse sorts.

We say, for instance, that Jane Austen wrote six novels; and in so saying we appear to imply the existence of the six novels that Jane Austen wrote. But just what are these novels? Austen wrote the novels by writing certain manuscripts. Yet the novels are not the author's manuscripts. A novel, at least once it has been published, does not cease to exist if the original manuscript is destroyed. One reads the novels by reading certain hardcovers or paperbacks. Yet the novels are not the hardcovers or paperbacks. If one borrows a paperback of *Emma*, reads it, buys a hardcover, and reads that, then one has read not two novels, but one novel twice. The novels, whatever they may be, are not made of paper and ink. They are not ordinary things: they are not things of the sort philosophers call **concrete**.

We say, again, that six is the number of Austen's novels, and that six is a perfect number. In so saying we appear to presuppose the existence of something, namely, the number six. But what is it? Surely not any particular inscription of the Roman 'VI' or the Arabic '6': those numerals only denote the number. Surely not any specific sextet or hexad, either: they only exemplify it. Numbers are if anything further than novels from being ordinary things. They again are things of the sort philosophers call **abstract**.

Colloquial speech and commonsense thought are liberally sprinkled with assertions appearing to involve non-concrete entities of all sorts, from novels to numbers. The language and theories of the natural and social sciences are not sprinkled but saturated with assertions apparently involving those quintessentially abstract entities, numbers: the fine structure constant, the gross national product, and so on indefinitely. As for pure mathematics, every branch of the subject abounds in assertions explicitly labelled 'existence theorems', beginning with Euclid's Theorem on the existence of infinitely many prime numbers. All this means that there is a wide range of widely accepted theories, including many commonsense, most scientific, and virtually all mathematical theories, to which you as an imagined convert to nominalism apparently cannot give any credence. And yet you cannot simply discard these theories, either.

For they are not idle. They guide practice in daily life and in specialized technologies. Simply to repudiate every view or opinion that appears to involve abstract entities would have so radical and so negative an effect on practice that no philosopher could sustain it for very long, or reasonably demand that others do so. You cannot simply dismiss mathematics as if it were mythology on a par with the teachings of Mme Blavatsky or Dr Velikovsky. A geologist interested in earthquake prediction or oil prospecting had better steer clear of Blavatsky's tales about the sinking of lost continents and Velikovsky's lore about the deposition of hydrocarbons by passing comets; but no philosopher will urge that the geologist should also renounce plate tectonics, on the grounds that it involves mythological entities like numbers and functions. At least, no philosopher who wishes to be taken seriously will so urge. Such is the problem that confronts you as an imagined convert to nominalism.

It is also the problem that really confronted the founders of modern nominalism. Though there were precursors active in the 1920s and 1930s, notably Stanisław Leśniewski, the continuous history of modern nominalism begins in the 1940s and 1950s with the work of Nelson Goodman and W. V. Quine, to whom most of the credit (or blame) for introducing the issue must be assigned. Their joint paper (Goodman and Quine 1947) is perhaps the earliest still regularly cited in the current literature. It opens with an oracular pronouncement to the effect that it is a 'philosophical intuition that cannot be justified by appeal to anything more fundamental' that abstract entities are to be renounced and dismissed. This pronouncement placed the authors in the problematic situation we have been describing. Their solution was to present nominalism not as a negative, destructive thesis, limiting itself to critique of the apparent implications and presuppositions of everyday, scientific, and mathematical theories, but rather as a positive, reconstructive project, seeking accommodation through reconstrual or reinterpretation of those ways of speaking that appear to involve abstract entities, so as to render at least a large part of them compatible with overarching nominalistic scruples.

While theories appearing to involve abstract entities range from the informal lore of untutored common sense to abstruse theories in pure mathematics, Goodman and Quine made it their priority to reconstrue the kind of science in which mathematics is applied, and especially the kind of mathematics applied in science. (They seem to have been resigned to having to speak loosely when speaking of matters not ready for strictly scientific treatment, and prepared to rank esoteric pure mathematics as mere speculation until it finds scientific applications.) After some modest initial progress, the project of Goodman and Quine reached an impasse. But they have by now come to have many successors, who hope to succeed where they failed in the nominalistic reconstrual of mathematically formulated science and scientifically applicable mathematics. The stream of publications by later reconstructive nominalists began as a trickle in the 1960s, grew in the 1970s, and became a torrent by the 1980s.

It also became a stream with many cross-currents and counter-currents, since contemporary reconstructive nominalists differ among themselves both as to what the ends of a reconstructive project are, and as to what means a reconstructive project may use in getting to them. As to differences of goals and ends, reconstructive nominalists may be likened to those ecumenically minded thinkers who have suggested that religion can be made perfectly congenial to humanists by (re)interpreting religious language so that 'God' refers, not to a transcendent supernatural being, but to something more innocuous, such as the good in human beings or an immanent historical process of liberation and enlightenment. But there is a great difference between offering such a reinterpretation as a substitute for more traditional creeds in which humanists have lost faith and offering it as an exegesis of what the canonical scriptures have really meant all along, despite the appearances to the contrary that have misled the unsophisticated. Similarly, there is a great difference between two construals of nominalistic construal, and of what the aim of such a construal should be.

On what may be called the **revolutionary** conception, the goal is reconstruction or revision: the production of novel mathematical and scientific theories to replace current theories. Reconstrual or reinterpretation is taken to be a means towards the end of such reconstruction or revision. It is taken to be the production of novel theories by assigning novel meanings to the words of current theories. While in principle not the only conceivable means towards the end of producing novel theories, it is in practice the most convenient means for nominalists who will have to go on for some time living and working with non-, un-, or anti-nominalist colleagues, since it produces novel theories that are pronounced and spelled just like the current ones. This permits the nominalist to speak and write like everyone else while doing mathematics or science, and to explain while doing philosophy, 'I didn't really mean what I said; what I really meant was . . .' (here giving the reconstrual or reinterpretation).

On what may be called the **hermeneutic** conception, the claim is instead, 'All anyone really means—all the words really mean—is . . .' (here again giving the reconstrual or reinterpretation). Reconstrual or reinterpretation is taken to be an analysis of what really 'deep down' the words of current theories have meant all along, despite appearances 'on the surface'. It is taken to be a means to the end of substantiating the claim that nominalist disbelief in numbers and their ilk is in the fullest sense compatible with belief in current mathematics and science.

Our frequent insertion of the qualifier 'apparently' and its cognates in the exposition thus far has been in an effort to remain neutral as between the two conceptions, revolutionary and hermeneutic, until the contrast between them could be noted. Since continuing in this way would be tedious for authors and readers alike, we will henceforth systematically drop the qualification 'apparently' and ignore the hermeneutic position, except in certain sections explicitly devoted to discussing it. And those will come only at the end of the book (in section III.C.2), since the question whether reconstrual should be regarded as alteration and emendation or as analysis and exegesis is best postponed until we have completed nominalistic reconstructive projects before us.

As to differences of ways and means, in a solemn terminology given currency by Quine, a theory implying or presupposing that entities of a certain sort exist (respectively, that predicates of a certain kind are meaningful) is said to involve an **ontological commitment** to such entities (respectively, an **ideological commitment** to such predicates, or the notions expressed by them). In a derivative sense, the commitments of a theory are also considered commitments of any theorist who asserts or believes it. All nominalists by definition agree in being unwilling to undertake ontological commitments to any sort of abstract entities. But nominalists have differed among themselves over almost every other ontological and ideological issue, so that the apparatus one nominalist invokes in a reinterpretive strategy very often will be quite unacceptable to another nominalist.

For example, Goodman and Quine and some of their successors have accepted as legitimate concrete entities **conglomerates** or complexes made up of bits and pieces, odds and ends, from this, that, and the other body. (Goodman, in Goodman and Leonard (1940), was a pioneer in developing the theory of such entities.) Other nominalists have regarded arbitrary conglomerates, whose boundaries need not be marked in any natural physical way, as being, if not abstract then at least philosophically problematic in the same way as abstract entities. (The later position paper Goodman (1956) contains a section (§3) responding to criticism, and the foregoing complaint occurs there as an item (iv) high on the list.)

For another example, many later nominalists have accepted **modality**, the logical distinction between what is the case and what could have been the case. The original nominalists did not, as Quine has emphasized in a later retrospective (his reply to Charles Parsons in Hahn and Schilpp (1986)):

> ... [L]ong ago, Goodman and I got what we could in the way of mathematics ... on the basis of a nominalist ontology and without assuming an infinite universe. We could not get enough to satisfy us. But we would not for a moment have considered enlisting the aid of modalities. The cure would have been far worse than the disease.

Some later nominalists have not only claimed that modality is philosophically problematic in the same way as abstractness, but have even made hermeneutic claims to the effect that modal distinctions covertly involve certain exotic abstracta, unactualized possible entities inhabiting unactualized possible worlds.

For yet another example, one later nominalist has invoked **geometricalia**, points and regions of physical space, while many other later nominalists have held these to be abstract and as philosophically problematic as anything else that is abstract.

The aim of this book is to chart the main currents in the nominalist stream. After setting up a common framework in Chapter B of this part, a dozen or so strategies will be outlined, three at length in Part II, and the rest more briefly in Chapter III.A. The relationship between our somewhat idealized versions of the strategies and actual proposals in the literature will be traced in Chapter III.B. The aim will be to make precise just what apparatus each strategy requires and just how large a part of scientifically applicable mathematics and mathematically formulated science each strategy can accommodate, thus facilitating comparison of different proposals as to how to preserve mathematics as a distinctive subject while abolishing its distinctive objects.

Inevitably, the issues here are often somewhat technical, but formal prerequisites are modest. Knowledge of introductory-level logic—essentially just the ability to read formulas written in symbolic notation—must be assumed. Acquaintance at least with some semi-popular account of intermediate-level logic would be helpful, but is not indispensable. The few discussions of issues of logic that may seem abstruse to readers without such background will be relegated to separate sections that may be considered optional semi-technical appendices.

Our hope is that after technical matters have been dealt with in a systematic and uniform way, the ground will have been cleared for a more properly philosophical evaluation of the pros and cons of nominalism. But no decisive evaluation of reconstructive nominalism will be attempted here. Only a very provisional philosophical evaluation will be attempted in our Conclusion, Chapter III.C.

In principle, at this point we as authors could have proceeded directly to the examination of the various strategies (and you as reader still can so proceed, by skipping ahead to the next chapter). In practice, at every point where a philosophically controversial move is made, wherever philosophical objections to the apparatus deployed or strategy pursued in some reconstructive project are raised, it will be necessary to refer back to questions about the motivation for engaging in a reconstructive project in the first place. For this reason, and especially because questions of motivation for engaging in a project are obviously logically prior to questions about the manner of execution of the project, the remainder of this Introduction will be devoted to addressing underlying motivations.

b. . . . of This Chapter

Prior to the question of how the nominalist is to go about reconstruing mathematics and science comes the question whether the nominalist needs to do so at all. In science, to be sure, when theorists become convinced that a scientific theory, despite successful practical applications, cannot be right because it conflicts with important scientific principles, they generally do undertake to develop a new theory conforming to those principles, one that will among other things explain the past utility of the old theory (and perhaps justify its continued use in certain contexts). Accordingly, when Albert Einstein became dissatisfied with classical gravitational theory, because despite its many important successes it was incompatible with the principle of relativity, he undertook to develop a new and relativistic gravitational theory, and so initiated a major revolution in science. Likewise, when Karl Weierstrass became dissatisfied with classical mathematical analysis, because despite its many important applications it did not conform to the ideal of rigour, he undertook to develop a new and rigorous version of mathematical analysis, and so instituted a major reform in mathematics.

But in philosophy, by contrast, when theorists become convinced that a scientific theory, despite successful practical applications, cannot be right because it conflicts with important philosophical principles, they do not always undertake to seek a new theory, by reinterpretation of the old theory or otherwise. Often they are content to insist that the current theory is a fiction, to concede that it is a very useful one, and to do nothing more. Or rather, they do nothing more in the direction of actively seeking some alternative theory that would explain why the theory they disbelieve is so useful in practice, beyond the 'theory' consisting of the bare assertion that the world behaves more or less as if the theory they disbelieve were true. They do nothing more than quietly cultivate an attitude of detachment, sometimes called **instrumentalism** (or **fictionalism** or even **utilitarianism**, though all these labels also have other uses). If other nominalists are like those twentieth-century humanists who want to de-supernaturalize religion, an instrumentalist nominalist might be likened to one of those free-thinkers of earlier times who, taking organized religion to be a crucial prop to the social order, publicly professed the established creed, while privately regarding it as superstition. Prior to the question how a nominalist's positive, reconstructive project should proceed comes the question whether a nominalist might not just be content to be such an instrumentalistic free-thinker.

And prior to that question comes the question why one should be a nominalist in the first place. Nominalists tend to call the opposing view **realism**, or more often **platonism** or **Platonism**, thus hinting that there is something mystical about it, as there historically was about Platonic and especially neo-Platonic philosophy. But for many of us, a kind of minimal non- or un- or anti-nominalism is simply our starting-point before we come to philosophy.

Before we come to philosophy, we have a fairly uncritical attitude towards, for instance, standard results of mathematics, or such of them as we have learned about. Having studied Euclid's Theorem, we are prepared to say that there exist infinitely many prime numbers. Moreover, when we say so, we say so without conscious mental reservation or purpose of evasion. We have in mind no subtle and complex attitude combining outward feigning with inward detachment. We do, to some extent, have such an attitude towards elementary physics, since our teachers will have told us that Newton's and Maxwell's theories, which we have studied, are only approximations to more complicated and more correct theories. But our teachers of elementary number theory will have told us nothing comparable.

Nor will they have told us that talk of numbers should be regarded as just a manner of speaking, in the way that our teachers of biology will have warned us that teleological language, talk of function and purpose, is a mere *façon de parler*, a shorthand for a longer and truer description in terms of evolution by natural selection. We will not have in mind, as we repeat Euclid's Theorem, any sophisticated reinterpretation of what we say as a mere figure of speech. At the same time, while our positive

conception of the nature of the numbers in whose existence we thus acquiesce may be of the haziest, we at least understand that numbers are not
supposed to be like ordinary concrete things like rocks or trees or people.
We do not expect to bump into them in the street. For that matter, we
understand as well that they're also not supposed to be like extraordinary
concrete things such as neutrinos or neutron stars. We don't expect them
to be detected using particle accelerators or radio telescopes. In this sense,
we acquiesce not only in their existence, but also in their abstractness.

For those of us for whom something like this is the starting-point, any
form of nominalism will have to be revisionary, and any revision demands
motivation. We are not so firmly attached to our pre-philosophical beliefs
that we would refuse to give them up even if an angel came down from
heaven and told us they were false; but if it is only a matter of a philosopher coming out of the study to tell us they are untrue, we will want
to be given some reasons for changing our minds. Why not just acquiesce
in the minimal non- or un-nominalism many of us find ourselves coming
to philosophy with?

And prior to that question comes the question just what nominalism is
supposed to be. We have said a nominalist is one who denies that abstract
entities exist, as an atheist is one who denies that God exists. But just
what does this mean? The Pythia, who spoke for Apollo at Delphi, was
notorious for her obscurity. The legendary Sibyls had likewise a reputation for murkiness. In line with this tradition, there is a certain lack of
clarity also in the vatic utterance of the founders of nominalism that all's
concrete, naught's abstract. Our trouble in understanding this revelation
is not about the form of what is asserted and denied, about the notion of
existence. Our question is, rather, about the notions of abstractness and
concreteness: just what is the content of distinction between them? What
makes something an abstractum rather than a concretum?

Our discussion thus far has operated with a rather vague and inchoate
understanding of what abstractness consists in. So will the discussion of
various reconstructive strategies in the central chapters of this book, for
so does the discussion in most of the literature to be surveyed in those
chapters, the literature on ways and means of nominalistic reconstruction.
But when one turns to discussion of reasons and grounds for purging our
views of all commitment to abstract entities, one will need to appeal to
some explicit account of what it is about abstract entities as such that
is supposed to make them philosophically problematic. If there is to be
serious discussion of the case for nominalism, something must be said
about what is supposed to be distinctive about the things the nominalist

is concerned to do without. This question in an obvious sense has the highest priority of all.

The questions we have identified, in order of logical priority, are: (i) What is an abstract entity? (ii) Why should one disbelieve in abstract entities? (iii) Why should one who disbelieves in abstract entities seek to reconstrue theories that involve them? These questions will be taken up in the order listed in sections 1–3. Since our aim in this book is mainly to clarify the technical side of recent wo. on nominalism, our discussion in the remainder of this Introduction of the foregoing questions on the motivational side will be somewhat in the nature of an extended digression. Given the limited space we will be devoting to that discussion, we can hardly hope to settle any major philosophical question, and we will in fact be aiming more to identify some philosophical issues calling for further reflection than to settle any philosophical issues currently in dispute. Given that aim, our discussion will tend to emphasize points that seem to us to cast doubt on what we take to be the plurality, if not the majority, view on questions (i)–(iii).

For while it would be hyperbole to speak of the existence of a general consensus or received view or conventional wisdom on such questions, certain opinions do seem to be very widespread indeed, not just among philosophers who have declared for nominalism in their work, but among philosophers generally. The single most common opinion on each of the foregoing questions we take to be somewhat as follows: (i) the question whether abstract entities exist is a rather exceptional case of a philosophical question where no extended preliminary analysis of the terms in which the question is posed is needed before launching into attempts to find an answer; (ii) there is a rather obvious prima-facie case for the negative answer to that question, a case for nominalism that initially seems quite powerful; (iii) however, nominalism pretty clearly must be judged untenable unless an appropriate reconstruction or reconstrual of science in conformity with its tenets can be developed, which initially seems quite difficult.

As a result of the prevalence of such opinions, discussions of nominalism more often than not begin with technical questions of the kind the body of this book will seek to clarify, questions about the feasibility of one or another reconstructive programme or project, passing over with only very brief remarks the questions we have identified as prior. If we accomplish nothing else in the remainder of this Introduction, we hope to raise some serious doubts as to whether that is indeed the appropriate place to *begin* the discussion.

1. WHAT IS NOMINALISM?

a. Paradigms of Abstractness

The distinction between abstract and concrete has a curious status in contemporary philosophy. Everyone concedes that it is not an ordinary, everyday distinction, with a consistent use outside philosophy. Almost no one troubles, however, to explain it at any length. David Lewis has examined such brief explanations as have been given in the literature, has classified them under four heads or 'Ways', and has published his results as a digression in one of his books (Lewis 1986: §1.7).

There can hardly be doubt as to which of the Ways is most popular. It is the 'Way of Example', which introduces the notion of abstractness by a short list of paradigm cases. That was the procedure of Goodman and Quine at the beginning of their joint paper: 'We do not believe in abstract entities. No one supposes that abstract entities—properties, relations, classes, etc.—exist in spacetime; but we mean more than this. We renounce them altogether.' That has been the procedure of most of the other nominalists whose work we will be surveying. To quote just the first of them, we read on the opening page of Field (1980):

Nominalism is the doctrine that there are no abstract entities. The term 'abstract entity' may not be entirely clear, but one thing that does seem clear is that such alleged entities as numbers, functions and sets are abstract—that is, they would be abstract if they existed. In defending nominalism I am denying that numbers, functions, sets or any similar entities exist.

The introduction of the distinction between abstract and concrete in most nominalist works tends to be thus casual—almost as much so as our own introduction of the distinction at the beginning of this chapter. Reading such passages as those quoted, with their short lists of examples, followed by 'and so on', one might think that the authors are supposing that their readers only need to be reminded of a distinction with which they are already familiar from some generally received standard explanation, such as might be found in elementary expositions to which the uninitiated could refer. In fact, however, there is no such generally received account.

But this does not mean that the distinction between abstract and concrete has no content. For the striking thing is that this procedure of definition by example has led to a fairly broad consensus—if not among all philosophers who use the terminology, or all those considered by Lewis, at least among the major contributors to the literature on reconstructive nominalism to be surveyed in this book, and most of the principal critics and commentators on their work—as to which sorts of objects count as abstract, and which as concrete. A longish and fairly generally agreed list of paradigms of abstractness (and of foils to these, paradigms of concreteness) can be drawn up.

Some abstracta come higher up on the list than others, not in the sense of being *more* abstract than other abstracta, since distinction of abstract and concrete is one of kind and not degree, but rather in the sense of being more *paradigmatically* abstract. At the very top of the list of abstracta come mathematicalia. These include natural, real, and other numbers. They also include sets or classes, also known as extensions or **collections** (in contrast to **individuals**). Mathematicalia may be divided into the **pure**, including the number two and the unit set of the null set, and the **impure**, including the number-with-units two metres and the unit set of any concrete entity.

Also high on the list of abstracta are metaphysicalia, abstract entities postulated in metaphysical speculation. Among the tamer metaphysical entities, for which the metaphysical theories about them can plausibly be claimed to be simply elaborations of commonsense thought, are properties and relations, known together as **universals** (in contrast to **particulars**). Among wilder metaphysical entities are **possibilia**, unactualized possible worlds and the unactualized possible entities that inhabit them.

Further down on the list of abstracta come **characters**, as we call them; for there is no generally agreed label. Here entities that are equivalent in some way are said thereby to have in common the same character, with different ways of being equivalent corresponding to different sorts of characters. One example is provided by biological species: organisms that are biologically conspecific or species-mates thereby have in common their species. Another example is provided by geometric shapes: figures that are geometrically similar or like-shaped thereby have in common their shape.

Under the heading of characters come important linguistic entities. The main objects of the branch of linguistics known as **semantics**, intensions or meanings, are what expressions that are synonymous or like-meaning thereby have in common. The main objects of the branch of linguistics known as **syntax** are expression **types** as contrasted with expression **tokens**. When one says that there are hundreds of letters on this page, one is speaking of tokens, concrete inscriptions in ink; when one says there are twenty-six letters altogether, one is speaking of types, abstract patterns of inscription, closely akin to shapes. The distinction between types and tokens applies to words, sentences, and books as well.

At the bottom of the list of abstracta come some miscellaneous, unclassifiable items, notably institutions. These include both informal ones, established by custom, and formal ones, established by law. An institution is not to be identified either with its physical facilities and equipment or its human participants and personnel. Yet institutions are so closely tied to such concreta as to be certainly atypical and arguably borderline as examples of abstracta. Still, they are not quite an outright disputed case, as are geometricalia.

At the head of the list of concreta come physicalia. These may be divided into physical objects, occupying space, and physical events, occurring in time. Only the former will be discussed explicitly here. At the head of the list of physical objects come observable ones, including all ordinary material bodies, animal or vegetable or mineral, natural or artificial. Next come extraordinary theoretical entities, from quarks to black holes, posited by physicists, inferred rather than directly observed. Further down the list of concreta come physical entities posited less by physicists than by metaphysicians. These extraordinary entities include absolutely arbitrary parts of ordinary bodies, and absolutely arbitrary wholes made up from different such parts of different such bodies, which we referred to earlier as conglomerates: the material contents, however mixed and heterogeneous, of any region of space, however irregular or disconnected, constitute a conglomerate.

Also near the top of the list of concreta are mentalia, minds and spirits, along with their contents, their perceptions and thoughts and volitions. Mentalia are themselves divided into those pertaining to embodied minds, as studied by psychology, and those pertaining to disincarnate spirits, as posited by theology.

It is doubtless unnecessary to say that properly speaking the foregoing list should not be described categorically as a list of things that are abstract and things that are concrete. Rather, it is a list of things that are so if they exist. Nominalists deny the existence of all the ones that count as abstract. Moreover, probably no single anti-nominalist believes in all the abstracta on the list. And probably no single philosopher, nominalist or anti-nominalist, believes in all the concreta. Each of the main categories of concreta has, in fact, been the topic of a debate older and more famous than the debate over nominalism. Thus George Berkeley denied the existence of perceptible physical entities as things distinct from perceptions of them, while Ernst Mach affirmed the existence of macroscopic, observable physical entities, but denied the existence of microscopic, theoretical ones.

Reconstructive nominalists generally agree in giving priority to reconstruing mathematically formulated science and scientifically applicable mathematics, so that the theories to be dealt with generally involve just observable and theoretical physical entities on the one hand, and abstract mathematical entities on the other. They generally agree that mathematical entities must be eliminated from such theories, and that in the course of eliminating them such entities as properties, possibilia, linguistic expression types, or the like must not be overtly or covertly introduced. They generally agree that whatever physical entities are accepted by common sense and current science may be retained: few if any reconstructive nominalists wish to go beyond nominalism to a more general **scepticism** (sometimes euphemistically called **empiricism**) of the type of Mach, let alone of the type of Berkeley. Many reconstructive nominalists do go beyond nominalism to a more general **materialism** (often euphemistically called **physicalism**), involving denial of minds and spirits. Other reconstructive nominalists accept mental entities in principle, but even they seem generally agreed that in practice the introduction of mental entities in the course of reinterpreting a physical theory would be inappropriate. Thus in practice reconstructive nominalists avoid controversy among themselves over many points that are controversial among other philosophers.

But they have not avoided controversy on all points (as noted already in article o.a), and while agreement on the classification of a list of examples and counter-examples is welcome, a list of uncontroversial cases cannot by itself decide these controversial cases.

b. Criteria of Abstractness

Even more importantly, when a proponent of nominalism seeks to argue that there is something wrong with abstracta, what is claimed to be wrong with them surely cannot be simply their occurrence on some list. Some account of what it is that the items on the list have in common is thus needed. In seeking further clarification of what distinguishes abstracta from concreta, one might turn from the recent literature to older discussions of abstraction and related matters; or instead one might try to relate the abstract/concrete distinction to other distinctions, presumed better understood; or else one might examine the brief, scattered remarks of recent writers that directly address the problem of characterizing abstractness, especially those that are mentioned in the brief, scattered passages where recent writers indicate what they take the grounds for nominalism to be.

These three routes roughly correspond to the three additional Ways of Lewis (beyond his Way of Example). First, there is the Way of Abstraction:

Abstract entities are abstractions from concrete entities. They result somehow from subtracting specificity, so that an incomplete description of the original entity would be a complete description of the abstraction.

Second, the Way of Conflation:

The distinction between concrete and abstract entities is just the distinction between individuals and collections, or between particulars and universals, or perhaps between particular individuals and everything else.

Third, the Way of Negation:

Abstract entities have no spatio-temporal location; they do not enter into causal interaction.

Lewis finds these Ways to conflict rather seriously with each other and with the list provided by the Way of Example, and doubts whether there is a coherent concept of abstractness. We find, however, that—at least for the range of authors with whom we will be concerned—there seems to be strong enough agreement on an open-ended enough list of examples and counter-examples to suggest that there is some coherent principle of classification implicitly at work. The task of articulating explicitly the implicit principle of classification underlying a list of examples and counter-examples—or **conceptual analysis,** as it is called—is notoriously difficult. It should therefore not be surprising or disturbing if success is achieved only after many failures; if quite a few ways that turn out to be dead ends have to be explored before the way that leads us where we want to go is found.

The Way of Abstraction might be called the Way of History, since it in effect directs attention to eighteenth- and nineteenth-century notions of abstraction, which were much debated by philosophers and mathematicians alike, from George Berkeley to Georg Cantor. On closer inspection, however, it will be found that the eighteenth-century debate in philosophy was over abstract *ideas* (and the older discussion in grammar was over abstract *nouns*). Thus it was mental (or linguistic) *representations* that were classified as abstract or otherwise, and not the entities represented. As for the nineteenth-century debate, it was over abstraction as a mental process, and abstractions as the products of such a process. For instance, Cantor's claim was that by a process of selective inattention of more or less the kind described by Lewis in the Way of Abstraction—a process of ignoring all aspects that distinguish the elements of a set from one another except their order—the mind creates the 'order type' or 'ordinal number' of the set.

Thus the discussion was infected with a confusion repudiated by virtually all contemporary philosophers, nominalist and anti-nominalist alike, namely, with **psychologism**, the failure to distinguish numbers themselves (which are abstract in the contemporary sense) from mental ideas or thoughts of them occurring to particular people at particular times and places with particular causes and effects (which are concrete). The current usage of 'abstract' by philosophers who are almost uniformly antipsychologistic is thus not a straightforward continuation of the earlier usage. Although there are scattered anticipations in earlier writers, the contemporary use of the term hardly dates back before Quine's early papers on ontology. At the beginning of one of his works, Quine (1951), he explains his 'ethics of terminology'. Terms that always were meaningless or that have fallen into desuetude he feels free to assign a new meaning. Hence his new usages of 'ontological' and 'ideological'. The current usage of 'abstract' should probably be viewed as of a piece with these, except that Quine provides no snappy formula by way of definition.

The case is similar with the label 'nominalism'. Contemporary writers often allude to the legendary William of Occam, leader of the legendary medieval nominalists, but no continuous tradition links present-day writers with the historical William Ockham and the historical fourteenth-century *nominales* who are the subjects of scholarly studies like Adams (1982) or Normore (1987). The present-day usage of 'nominalism' hardly dates back beyond the work of Goodman in the 1940s or thereabouts. The application of the medieval term 'nominalism' to his modern view did seem historically appropriate to him in the light of the understanding of the Middle Ages available to him as a non-specialist at that time. But when it was complained that his usage was something of a misnomer (item (i) on the list of objections in his position paper (Goodman 1956) mentioned earlier) he made no very strong claims to historical appropriateness: 'I claim no more than that the principle I have set forth is one reasonable formulation of the traditional injunction [the so-called **Occam's Razor**] against undue multiplication of entities.' Our impression as non-specialists is that subsequent historical studies tend to make the connection between medieval and modern nominalism seem ever more tenuous. (For a bit more on this history, see Rosen 1992: chapter 1.)

Setting aside history, it is possible to construe the Way of Abstraction somewhat differently. One may take the suggestion to be, not that abstracta *are* the products of a mental process of selective inattention, but that they are the kinds of objects that psychologistically inclined philosophers of earlier times erroneously *took to be* such products. This would come close to identifying abstracta with characters, for what incomplete description does in the way of subtracting specificity is precisely to ignore whatever distinguishes a given entity from any equivalent one. And identification of abstracta with characters would in effect be another version of the Way of Conflation, alongside the identification of abstracta with properties, or with sets.

All versions of this Way have in common that they assimilate the category of abstracta to some other category supposed to be better understood. As the use of the pejorative term 'conflation' for this assimilation suggests, every version of this Way faces serious objections. The objections are perhaps sufficiently illustrated by the case of the version identifying abstracta with sets.

A philosopher who was not willing to go all the way with the nominalists might well wish nonetheless to reduce the vast variety of categories of abstracta involved in scientific theorizing and commonsense thought, by choosing some one privileged category and seeking surrogates or proxies for abstracta of other categories within this privileged one. A nominalist philosopher, too, might well wish to proceed in two stages when eliminating abstracta, first eliminating all other categories of abstracta in favour of some one distinguished category, and then eliminating that distinguished category also. For philosophers of either kind, sets or classes would be an inviting choice of privileged or distinguished category. For set-theoretic surrogates are available for virtually all mathematicalia (as will be explained in article B.1.a), and set-theoretic proxies immediately suggest themselves for characters and properties as well: a convenient stand-in for the character that a given entity has in common with all entities equivalent to it would be the set of all entities equivalent to the given one; an obvious *locum tenens* for a property would be the set of all particulars enjoying it.

It is one thing, however, to claim that any other abstractum can be eliminated in favour of an ersatz set-theoretic entity, and quite another to claim by way of clarification of the notion of abstractness that to be an abstractum just means to *be* a set. Such a claim is hardly plausible—and is seldom made—even in the case of the reduction of numbers to sets. For instance (as will be mentioned in Chapter II.A), the set-theoretic surrogates for real numbers were more or less self-consciously put forward as *replacements* for the real numbers as traditionally conceived, the traditional (geometrical) conception having come to be considered untenable (after the development of non-Euclidean geometry).

If such conflation is rejected, there remains only the route that attempts a direct characterization of the features the possession of which make for abstractness. This is a *via negativa*, the Way of Negation, since the features most often cited are negative ones: (i) lack of spatial location; (ii) lack of temporal location; (iii) causal impassivity; (iv) causal inactivity.

The question is whether there is some one, or some combination of several, of these features (i)–(iv) that is enjoyed by all abstracta and no concreta, and so would be usable as a criterion for distinguishing abstract from concrete. In checking against the list of paradigms and foils, arguably no great weight should be put on the wilder and more exotic examples from physics or metaphysics, such as quantum wavicles or possible worlds. A rough criterion that at least gets the tamer cases right can always be refined once found. Certainly refinement can be expected to be called for, since esoteric physics often puts pressure on commonsense notions of space and time, and cause and effect, and esoteric metaphysics sometimes puts pressure even on commonsense notions of logic.

To begin at the head of the list of paradigms, it does seem that pure mathematicalia, such as natural numbers, exhibit all four features above. As for (i), numbers are not normally thought of as existing at any place in space. Sometimes it is said that they exist 'outside' space, but the connotations of the preposition make this formulation misleading. It may subconsciously suggest that while numbers are not located at any ordinary place, they are located at some extraordinary place 'beyond' all ordinary ones. In fact, predicates of space or place are simply inapplicable to numbers. It would betray a misunderstanding for someone presented with a proof of the existence of infinitely many primes to ask, 'Where do these primes exist? At what place may they be found?'

As for (ii), one cannot quite say that numbers do not exist at any moment in time. For it is a peculiarity (or defect) of English that, unlike some natural and most artificial languages, it obliges one to put every verb in some tense or other and thus to locate everything spoken of in the past, present, or future of the moment of speech. Thus to say that something does not exist at any moment in time is to imply that it does not exist at all. What one *can* say is that numbers do not exist at any one moment rather than another: if they ever did or do or ever will exist, then they always have existed and exist and are always going to exist. It is in this sense that it would betray a confusion to ask, 'When did (or when will) these primes exist? At what time may they be found?'

As for (iii) and (iv), it would likewise be indicative of muddleheadedness to ask, 'What is the cause of the existence of these primes? Who made them? What is the effect of the existence of these primes? What do they do?' Numbers are not agents or patients; they make nothing and do nothing; they are made by nothing and suffer nothing. This, at least, is the received view: virtually no one who holds that they exist holds that they could perfectly well have failed to exist, so that one can meaningfully ask about the causes why they happen to exist; and virtually no one who holds that they do not exist holds that they could perfectly well have existed, so that one can meaningfully ask about the causes why they happen not to exist.

Turning to items lower on the list of paradigms than numbers, we encounter difficulties. These can be brought out by considering a fantasy of the kind in which philosophers like to indulge. Imagine that in some remote galaxy at some remote future time there will exist a perfect double of our home planet, a Duplicate Earth. There one will find duplicates of every concretum under the sun. But one will find no duplicate of the number three, for instance: trios and troikas and threesomes on Duplicate Earth will be simply more instances of the very same number three as is instanced by triads and triplets and trinities on Earth. But what about other abstracta?

The woolly creatures grazing in the meadows of Duplicate Earth being products of convergent evolution, having no common ancestry with earthly sheep, it will be tempting to say that they will belong, not to the species *Ovis aries*, but to a duplicate thereof, *Ovis aries bis*. Likewise, it is tempting to say that the duplicate Jane Austen will not be a belated co-author with the original Jane Austen of *Mansfield Park*, but rather will be the author of a duplicate novel, *Mansfield Park Redux*. As for institutions, it is more than tempting to say that the IBM Corporation is multinational but not polygalactic, and that it will not have more factories, offices, and outlets, more employees, managers, and customers on Duplicate Earth, but rather that on Duplicate Earth there will be a duplicate corporation, IBM II.

It now seems natural to say that an earthly species or novel or corporation is located on Earth, and its *Doppelgänger* on Duplicate Earth, contrary to (i); and that the earthly examples exist now, but the duplicates only at a remote later period, contrary to (ii). Or rather, it seems natural to say that the duplicates don't and won't exist, since Duplicate Earth doesn't and won't, but that they would have if it had; and similarly, that the originals exist only because Earth does, and wouldn't have if it hadn't. Moreover, it seems natural to say of almost anything whose existence is thus temporally and spatially circumscribed and is thus contingent that there must be some cause of its existing, and of its existing when and where and how it does, contrary to (iii); and presumably also some effects of its existing and of its existing when and where and how it does, contrary to (iv).

Most of these points can be made even without the aid to intuition provided by the fantasy. Thus species are assigned to geographical ranges and geological epochs, and their existence is explained in terms of environmental and other causes, and cited in turn as the cause of environmental and other effects. Also novels are classified by country and period, and their existence is considered to be the work of their authors, and an influence on their readers. Some categories of abstracta may be spoken of colloquially not only in causal but even in perceptual terms: an ornithologist may be said to discern the shape of the silhouette of a flying bird against the sky, and thereby detect its species.

Nothing so far has been said about sets of concrete individuals or properties of concrete particulars. Metaphysical theories about the former are put forward in Maddy (1990*a*), and of the latter in Armstrong (1978), with the authors in both cases claiming that most of what they say is quite in accord with, or at any rate a fairly natural elaboration of, what is said in ordinary or scientific language and supposed by common sense or scientific thought. Given this work—which cannot be adequately summarized in the space available here—there seems to be much to be said for the claims that collections of concreta are present where and when their individual elements are, and act (and are acted upon) when their elements act (or are acted upon) collectively; likewise, properties of concreta are present where and when their particular instances are, and act and are acted upon when their instances act or are acted upon appropriately (meaning, on account of their having the property in question).

The concession is usually made, however, that sets and universals are not located, and do not interact, *in quite the same way as* ordinary material bodies. This suggests reformulating criteria (i)–(iv), adding in each case the qualifying clause, 'at least not in quite the same way . . .'. The task, then, would be to try to articulate just what way this is.

Beginning with (i), it seems possible to say something fairly definite about the manner in which ordinary material bodies are located in space: each is in one place at one time, and no two are in the same place at the same time; when any of them occupies an extended region, it does so by having parts that occupy the parts of the region. By contrast, according to metaphysicians, the property of horsiness is (wholly) present wherever a horse is present, and since many different horses are present in many different places, this property is also (wholly) present in many different places at the same time. Again, according to the metaphysicians, the region occupied by the set of all horses is just the region occupied by all horses considered together (that is, by the mass of all horse-flesh). One part of this region is the region occupied by all horse-heads. This is the region occupied by the set of all horse-heads, but that set presumably is not a part of the set of all horses in the way that, say, the set of all race-horses is. Thus the manner of location of properties of concrete particulars and sets of concrete individuals does seem to differ from the manner of location of the particular individuals themselves.

It would remain to consider the mode of location of other sorts of abstracta, beginning with characters, but there is a more pressing problem coming from the side of concreta, the problem raised by the case of minds. For they on the one hand are supposed to be concrete, but on the other hand are not located spatially, at least not in the same manner as material bodies. Thus the criterion of spatial location in that manner does not seem to distinguish abstract from concrete correctly. Moreover, God is (at least according to one influential theological tradition) on the one hand concrete (and immanently and omnipresently and omnitemporally active), but on the other transcendent, eternal, and impassive. The conclusion to which these examples point is that (iv) rather than (i), causal activity rather than spatial location, must be taken to be the distinguishing note of the concreta, if any of the features considered so far can be.

Perhaps too much weight should not be put on these examples, given how many nominalists are willing to go beyond nominalism to a more general physicalism. Perhaps minds and God should be set aside along with the quantum wavicles and possible worlds excluded from consideration earlier. But the conclusion to which the examples pointed is borne out by other considerations: in arguing about controversial cases, and more importantly in arguing for nominalism, the complaint about abstracta is most often of their inactivity, inertness, and inefficacy.

The question before us thus becomes whether it is possible to clarify the distinctive way in which ordinary material bodies are causally active, and if so whether it can indeed be said that the various sorts of abstracta that are colloquially spoken of in causal terms are not causally active in that way. More precisely, the question should be whether it is possible to clarify the distinctive way in which ordinary material bodies participate causally in events. For if there is one comparatively uncontroversial point in the generally contentious area of the conceptual analysis of causation, it is that what primarily are related as cause and effect are events, with bodies being considered causally active in so far as they appropriately participate in events. Now the way in which ordinary material bodies participate causally in events is not so easy to characterize as was the way in which they are spatially located. For indeed, there is no one such way, but rather several ways.

These can be brought out by considering the anecdote reported by James Boswell, according to whom Samuel Johnson once attempted to refute Berkeley's philosophy by kicking a stone. Johnson, in kicking the stone, proved his causal efficacy. Had he sent the stone flying so that it toppled a nearby urn, the stone, too, would have proved its causal efficacy, though there can be no question of voluntary action in its case. As it was, Boswell reports that the stone proved its causal efficacy in a different way, by resisting the mighty force of the kick, so that Johnson rebounded from it.

Note, now, the following differences in manner of participation. Johnson's participation in the event of Johnson's kicking the stone involved an intrinsic change in him. Johnson after was no exact duplicate of Johnson before, since among other things the disposition of his parts, of his limbs, was different. But when a stone participates in the event of the urn being toppled, intrinsic changes in the stone, or for that matter the urn, are less important than changes in the external, spatial relations between them, which cannot be considered internal changes in anything, unless it were in the complex or conglomerate of stone plus urn. In the case of the stone's participating in the event of Johnson rebounding, it was the omission or failure on the part of the stone to change when it could have, namely, its omission or failure to yield or shift, that seems most pertinent.

So ordinary material bodies can participate efficaciously in events in several different ways, and the question is whether one can say that the various sorts of abstracta that are colloquially spoken of in causal terms do not participate efficaciously in events in *any* of these ways. In considering that question, let us begin with properties. The anecdote invites, more specifically, consideration of qualities of mind. For it was forthright impatience with philosophical sophistry, according to some, or ineradicable obtuseness towards philosophical distinctions, according to others, that caused Johnson to kick the stone. Here indeed one is inclined to say that though properties such as impatience or obtuseness are thus spoken of colloquially in causal terms, nonetheless in the event of Johnson's becoming

impatient with (or remaining obtuse to) Berkeley's philosophy there is
no change in the intrinsic character of any general quality of mind such
as impatience or obtuseness, nor any such change in any complex of
such qualities, nor yet any omission or failure to undergo change where
change could have been undergone. Thus the sense or manner in which
the properties may be spoken of as causes does seem to differ from
all the senses or manners in which ordinary material bodies may be so
spoken of.

If we turn next to characters, however, the results are less satisfactory.
For one is inclined to say that species, for instance, do undergo changes
in their intrinsic nature as well as in their external relations to other
species when they adapt to their environments, and that these changes
have definite causes and effects. Likewise book-types, from Boswell's
biography to Austen's novels, seem to undergo intrinsic changes as their
authors compose them, adding and deleting episodes.

Contemplating these and other complications, one may be overcome
with Johnsonian impatience and exclaim, 'Enough of such sophistry!
Physics tells us how ordinary material bodies act causally. They act by
exerting forces of one of four kinds: gravitational, electromagnetic, or
weak or strong nuclear. Biographies and novels, species and genera, exert
no such forces over and above that of the aggregate of the concrete tokens
and organisms pertaining to them. So they do not act causally in the
relevant sense—and there's an end on't!'

This proposal, however, though perhaps not exactly obtuse, misses the
point. What was wanted was more an analysis of the meaning of the
question, 'What constitutes physical action?' than a report of the answer
to the question given by current physics. As a result, the proposal makes
the distinction between abstract and concrete depend on what the actual
physical laws of the actual physical universe are, and so tends to make the
distinction inapplicable under counterfactual hypotheses not known to be
compatible with those laws. And as a result, the proposal tends to make
the distinction inapplicable to all the many nominalistic strategies that
involve consideration of what would have been the case under such counterfactual hypotheses (such as the hypothesis of the coexistence of infinitely
many non-overlapping ordinary material bodies).

Having now noted the knottiness of the problem of characterizing
abstractness, in line with our general aims in this book (as explained at
the end of article o.b), we will make no further effort to cut through or
unravel the problem ourselves, but rather will leave further reflection on
it to the interested reader.

2. WHY NOMINALISM?

a. Stereotypes

The thesis of nominalism has had a curious history. It was announced without supporting argument as 'a philosophical intuition that cannot be justified by appeal to anything more fundamental'. It immediately met with a barrage of counter-arguments, so that almost all the varieties of anti-nominalism found in developed form in the recent literature can be found in embryonic form already in papers from *circa* 1950. Influential supporting arguments appeared in the literature only a quarter-century or so after the announcement of the thesis—and then the most influential arguments have been quite indirect. Let us elaborate on this last point.

There are in general several ways to challenge an assertion. If Lestrade suggests that Professor Moriarty is the culprit, Holmes may directly challenge the truth of this suggestion: 'Watson, Lestrade is quite wrong. Villain though he is, Moriarty can't have perpetrated *this* outrage, since while the crime was in progress in Paddington Station, Moriarty was strolling innocently along the Strand, under my constant observation.' A more extreme challenge would be to the very *meaningfulness* of the suggestion: 'Watson, Lestrade must be going senile. He told me Moriarty did away with the victim because she had discovered he was a bachelor bigamist, and then disposed of her body by feeding it to his pet bandersnatch. He seemed quite oblivious to the facts that a bigamist by definition can't be a bachelor, and that "bandersnatch" is a nonsense word from *Jabberwocky*.' A less direct challenge would leave the issue of the truth of the suggestion untouched, and only question whether one is in a position to *know* it is true: 'Watson, Lestrade is leaping to conclusions. His suspicion is based on nothing more than his analysis of a bit of cigar ash found at the scene. He's learned enough of my methods to identify it as coming from the kind of tobacco Moriarty uses, but if he'd read my monograph more carefully he'd have learned that this is one of the most common kinds, smoked not just by the Napoleon of crime, but by half of London.'

In the case of the nominalist challenge to the assertion that there are abstracta, a direct approach, making the direct, ontological counter-assertion that there are only concreta, and seeking to support that counter-assertion by direct, ontological, or **metaphysical** argument, presents certain difficulties. For in metaphysics, unlike physics, there is no large corpus of results and methods generally agreed upon. So someone who wishes to argue for a controversial conclusion Q will almost never be able to do so on the basis of principles already generally accepted, but will have to suggest some new principle P. If P is supported by appeal to 'intuition', opponents of Q can be expected to argue back as follows: 'Mistaking initial plausibility for ultimate certitude is the philosopher's besetting sin. Your "intuition" that P has some superficial plausibility, but that immediately evaporates when one realizes what P would imply, namely, Q.' If P is supported by induction from examples, opponents of Q can be expected to argue back as follows: 'Hasty over-generalization from too narrow a range of examples is the philosopher's chief vice. Your inductive inference to P from a range of examples not including Q is a typical case of such over-hasty generalization.' Perhaps more significantly, there is so little agreement in metaphysics that it usually happens that even many proponents of Q won't think that P is the principle that should be invoked in arguing for it.

Moreover, no one who questions the weaker metaphysical claim that it would be false to deny P is going to assent to the stronger **semantical** claim that it would be meaningless to do so. And if what ultimately is being claimed to be meaningless is not some newly introduced jargon but a long-standing usage in some well-established field of science, the following kind of objection may also be expected: 'There is a regular little industry of publishing translations of foreign works, dictionaries of technical terms, and so on, in this field of science. Clearly it is "meaningful" in the ordinary sense of the term, the sense that concerns translators, lexicographers, and other trained professionals concerned with meaning. It is your extraordinary sense of "meaningful" that may well not be meaningful.'

Now, direct metaphysical arguments (and even extreme semantical claims) have nonetheless sometimes been met with. Most notably, one of the founders of nominalism, Goodman, has taken this line (in his position paper, Goodman (1956), cited earlier, and elsewhere). He advances the metaphysical principle P that distinct entities must have distinct constituents, and on this basis argues for the ontological conclusion Q that collections cannot exist. For the set or class $\{\{a\}, \{a, b\}\}$ is supposed to be distinct from the set or class $\{\{b\}, \{a, b\}\}$, even though both ultimately are constituted from the same a and b. (If one takes seriously Goodman's occasional use of the phrase 'makes no real sense', he is actually making the stronger, semantical claim that to assert the existence of sets or classes, or to deny his metaphysical principle, would be not just false, but meaningless.)

However, this direct metaphysical argument has had little influence. Most anti-nominalists concede that the principle P would imply conclusion Q (though Lewis (1991) is in a sense an exception). But anti-nominalists have not granted even that the principle P has a superficial initial plausibility. Rather, because it goes beyond one of the very few widely accepted principles in metaphysics, the principle that distinct entities must have distinct features of *some kind or other*, it immediately excites suspicion. Nor have most anti-nominalists granted even that the principle P holds in all examples pertaining to concrete entities. Rather, various prima-facie counter-examples have been suggested, for instance that a statue and the lump of clay of which it is made are distinct entities with the same molecular and atomic constituents. Moreover, nominalists rarely have wished to rely on P, perhaps because the argument from P only works to give the conclusion Q that collections do not exist, and does not seem to work against all the various other categories of abstracta.

Instead, nominalists for the past couple of decades have much preferred to rely on **epistemological** arguments, whose conclusion is that even if abstracta do exist they might as well not, since they will be unknowable. The source of inspiration for such arguments is curious. Though neither has ever described himself as a nominalist, Paul Benacerraf and Hilary Putnam deserve almost as much credit (or blame) for the contemporary prominence of the issue of nominalism as do Goodman and Quine. This is partly owing to the influence of their anthology, Benacerraf and Putnam (1964, 1983), in giving wide circulation to a selection of key papers and in shaping through its editorial introduction how the problem is perceived. But the main influence of the two philosophers has been through their own publications, beginning with Benacerraf (1965) and Putnam (1967). Above all, while Putnam (1971) has done more than any other work to encourage dissatisfaction with any merely negative, destructive nominalism, Benacerraf (1973) has done more than any other work to encourage sympathy for nominalism as such—and this is despite the fact that Benacerraf does not himself advocate nominalism! Benacerraf's paper presents a puzzle about how knowledge of abstracta could be possible, which his nominalistic readers have appropriated and transformed into an argument that knowledge of abstracta is impossible.

A collection of passages from nominalistic writings citing and appropriating or adapting Benacerraf's work can be found in the opening section of Burgess (1990*b*). Here is a pastiche of several of them:

It is a mystery how we concrete beings can know abstracta ... Numbers, sets, and the other entities to which mathematics ... [appears] to be committed are things we cannot perceive directly; indeed, ... we stand in no causal, or otherwise empirically scrutible contact whatever with them. But then how do we have knowledge of them? ... [Since] there are no causal connections between the entities of the platonic realm and ourselves, . . . it seems as if to answer [such] questions we are going to have to postulate some aphysical connection, some mysterious mental grasping.

Few nominalist works since the early 1970s have lacked a passage with this flavour.

If we had to offer a formulation in our own words of a stereotypical nominalist position, we might put it something like this:

We nominalists hold that reality is a *cosmos*, a system connected by causal rela-tions and ordered by causal laws, containing entities ranging from the diverse inorganic creations and organic creatures that we daily observe and with which we daily interact, to the various unobservable causes of observable reactions that have been inferred by scientific theorists (and perhaps to the First Cause postulated by religious thinkers). Anti-nominalists hold that outside, above, and beyond all this [and here one gestures expansively to the circumambient universe] there is another reality, teeming with entities radically unlike concrete entities—and caus-ally wholly isolated from them. This amounts to an especially unattractive variety of supernaturalism, somewhat like Epicurean theology. Compared with more traditional creeds, it offers no promise of reward to the faithful, since between them and the other world in which anti-nominalists would have them believe there is a great gulf fixed; but it requires quite as much in the way of faith to provide evidence of things unseen. Surely anti-nominalists owe us a detailed explanation of how anything we do here can provide us with knowledge of what is going on over there, on the other side of the great gulf or great wall. However difficult it may be to formulate *precisely* what is wrong with anti-nominalism, one need only consider how anti-nominalists depict reality (flesh-and-blood subjects on one side, ethereal objects on the other, a causally impenetrable great wall in between) in order to see at once that something is wrong.

If one looks for an *argument* in all this, what one finds is a syllogism, or rather, an enthymeme. The conclusion is explicit: 'We can have no know-ledge of abstract entities.' So is the minor premiss: 'We have no causal connection with abstract entities.' What is left tacit is the major premiss, some kind of **causal theory of knowledge**. This syllogism is the basic epistemological argument for nominalism. It has numerous versions and variants, and we will be examining briefly three of the most influential in articles 2.b–2.d.

In attempting to understand and evaluate the arguments for a philo-sophical position, it is usually helpful to have in mind at least a rough, vague notion of what the opposing position being argued against would be like. In the case of arguments for nominalism, however, what it is important to bear in mind is that there is no such thing as *the* opposing position. In order to underscore this point, we will offer very rough and

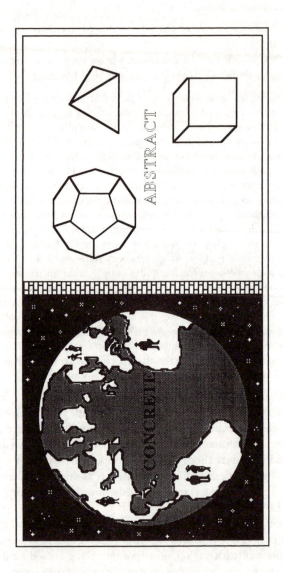

very brief descriptions of a few of the more important anti-nominalist positions.

Some anti-nominalists object more to the minor than to the major premiss of the nominalist syllogism. The self-described 'realist' Penelope Maddy is perhaps the best example. In Maddy (1990a) she acknowledges inspiration from the early anti-nominalism of Gödel (1947). Other positions not unrelated to hers will be cited later (at the end of article III.B.2.c). Roughly and briefly, her position has two aspects: on the one hand, it concedes that knowledge of entities of a given sort requires causal and even perceptual contact with at least some of them. On the other hand, it insists that we do have causal and even perceptual contact with at least some abstracta. This latter claim need not involve anything mysterious or occult, since we have already remarked (in article 1.b) that several sorts of abstracta are colloquially spoken of in causal and even perceptual terms. Indeed, perhaps the most important consequence of the difficulty of characterizing abstractness we remarked on earlier (in article 1.b) is that it leaves room for an anti-nominalist position of the kind being contemplated.

Most anti-nominalists object more to the major than to the minor premiss. The objections take two principal forms, both starting from the observation that nominalists are denying that certain entities 'really' exist, or that the belief that they do is 'really' justified, though that belief is well justified by ordinary commonsense and scientific and mathematical standards of justification. One form of anti-nominalist objection questions whether there is any meaningful notion of existence other than that constituted by ordinary standards of justification for existential beliefs. The self-described 'Fregean Platonism' of the St Andrews School of Crispin Wright and Bob Hale is perhaps the best example. In Wright (1983) and Hale (1987) they acknowledge inspiration from the early anti-nominalism of Dummett (1956). Not unrelated is the early anti-nominalism of Carnap (1950).

Roughly and briefly, the assumption behind this form of anti-nominalism is that meaning is constituted by rules of language that determine under what conditions a sentence counts as correct or justified. Some versions or variants write in this connection of **confirmation-conditions**, others of **verification-conditions**, yet others of **truth-conditions**; and these terminological differences mark significant doctrinal differences. The following description is merely of the least common denominator of the various views. For a more detailed discussion of the St Andrews School, see Rosen (1993a).

According to the rules of language, some sentences, called **analytic**, count as correct and justified under any conditions whatsoever; while other sentences, called **synthetic**, count as correct or justified only under certain conditions. In the case of an analytic sentence A, unless and until one has adopted the pertinent rules of language, the question whether it is the case that A has no meaning; but when and if one adopts the pertinent rules of language, then as soon as one does so, the question whether it is the case that A must receive an affirmative answer. The question cannot meaningfully be asked and answered in the negative. Such analytic truths are supposed to include among others logical, definitional, mathematical, and ontological truths: thus it cannot meaningfully be doubted whether numbers 'really' exist, according to this form of anti-nominalism, since we have no meaningful notion of the existence of numbers except that constituted by ordinary mathematical rules for when existence assertions about numbers count as correct or justified; and by *those* rules the general assertion of the existence of numbers does count as correct and justified (as do Euclid's Theorem and many other existence theorems).

Another form of objection questions whether there is any viable notion of 'justification' other than that constituted by ordinary commonsense and scientific and mathematical standards of justification. The best example of this form of anti-nominalism is provided by Quine. For Quine, after the failure of his joint project with Goodman, soon came to reconsider, and eventually came to recant, his nominalism. Perhaps the most concise and convenient expression of his ex-, post-, and anti-nominalist position is that in Quine (1966*b*). In Quine (1969) he presents his position as a response to certain historical developments in philosophy, which may be sketched in very quick strokes with a very broad brush as follows.

The seventeenth-century pioneers in the scientific study of nature tended to hold the view—which unlike the views of most present-day anti-nominalists can in a serious sense be called 'Platonist'—that the Creator designed the world mathematically. They tended to conclude that if scientists describe the world mathematically, then they may hope to achieve 'truth' in the robust sense of **correspondence** between human understanding and divine intention. The aim of epistemological meditations on first philosophy was to provide a philosophical foundation for science, showing how it can achieve truth in a robust sense.

The eighteenth-century forerunners of the scientific study of cognition tended to hold that our commonsense and scientific representation of the world is constructed by starting with the experiences available to creatures with sensory capabilities like ours, and then systematizing in what is the most simple or convenient (and in many cases the only possible or feasible) way for creatures with intellectual capabilities like ours: *we* can only effectively cope with what we perceive by representing it to ourselves as part of a system also having parts we do not perceive, and we can most efficiently cope with such physical systems by representing them to ourselves as approximate realizations of ideal mathematical systems. The thinkers in question tended to conclude that it is very doubtful whether intelligent creatures having cognitive capabilities differing from ours, to say nothing of an omniscient Creator, would need or have similar representations. Thus the result of epistemological enquiry concerning the human understanding was a critique of claims that common sense and science can achieve truth in any very robust sense.

Quine's response to all this is to abandon as futile if not meaningless the traditional **alienated** conception of epistemology, on which the epistemologist remains a foreigner to the scientific community, seeking to evaluate its methods and standards—a conception that presupposes other methods and standards of evaluation, outside and above and beyond those of science. (Along with it, he abandons any robust sense of truth.) In its place he advocates a novel **naturalized** conception of epistemology, on which the epistemologist becomes a citizen of the scientific community, seeking only to describe its methods and standards, even while adhering to them. (Along with it, he adopts the **disquotational** sense of truth, on which to assert that some theory is 'true' amounts to no more than asserting the theory itself.)

The naturalized epistemologist may largely accept the nominalist's description according to which our method of positing mathematical systems as models is just the most efficient way for us, with such cognitive capacities as we have, to cope with physical systems. For that matter, the naturalized epistemologist may largely accept the sceptic's description according to which our method of positing a physical system with parts we do not perceive is just the only effective way for us, with such cognitive capacities as we have, to cope with what we do perceive. What the naturalized epistemologist rejects is the suggestion that such descriptions of our methods show that those methods, though scientifically justified, which is to say justified by scientific standards, are not 'really' justified, which is to say not justified by 'real' standards. For the search for 'real' standards exterior and superior and ulterior to those of 'common sense and the refined common sense which is science' is pointless if not unintelligible.

Suspicion about the pretensions of philosophy to judge common sense and science from some higher and better and further standpoint comes to the foreground in some anti-nominalist positions not otherwise especially close to Quine's, and also forms an important part of the background even of anti-nominalist positions like Maddy's and Carnap's that bring other considerations to the fore. A particular forceful expression of such an attitude has been given by David Lewis (in the 'Credo' in Lewis 1991: §2.8, reiterated in Lewis 1993):

Renouncing classes means rejecting mathematics. That will not do. Mathematics is an established, going concern. Philosophy is as shaky as can be. To reject mathematics on philosophical grounds would be absurd. If we philosophers are sorely puzzled by the classes that constitute mathematical reality, that's our problem. We shouldn't expect mathematics to go away to make our life easier. Even if we reject mathematics gently—explaining how it can be a most useful fiction, 'good without being true'—we still reject it, and that's still absurd. Even if we hold on to some mutilated fragments of mathematics that can be reconstructed without classes, if we reject the bulk of mathematics that's still absurd. . . . I laugh to think how presumptuous it would be to reject mathematics for philosophical reasons. How would you like to go and tell the mathematicians that they must change their ways, and abjure countless errors, now that philosophy has discovered that there are no classes? Will you tell them, with a straight face, to follow philosophical argument wherever it leads? If they challenge your credentials, will you boast of philosophy's other great discoveries: that motion is impossible, that a being than which no greater can be conceived cannot be conceived not to exist, that it is unthinkable that anything exists outside the mind, . . . and so on, and on ad nauseam? Not me!

If we had to offer a stereotype of anti-nominalism to go with our earlier stereotype of nominalism, we would offer something in this general vicinity:

We come to philosophy believers in a large variety of mathematical and scientific theories—not to mention many deliverances of everyday common sense—that are up to their ears in suppositions about entities nothing like concrete bodies we can see or touch, from numbers to functions to sets, from species to genera to phyla, from shapes to books to languages, from games to corporations to universities. To be sure, we also come to philosophy in principle prepared to submit all our pre-philosophical beliefs to critical examination and to revise them if good reasons for doing so emerge. But we anti-nominalists hold the *onus probandi* to be on the advocates of revision; and in practice the historical record of philosophical and theological 'corrections' to science and mathematics, from Bellarmine's 'correction' of Galileo onwards, has been so dismal that we will demand very good reasons indeed from anyone who comes before us with another such philosophical claim of massive 'error' in science: appeals to 'philosophical intuition', generalizations from what holds for the entities with which we are most immediately familiar to what must hold for any entity whatsoever, and similar modes of argumentation will not suffice to induce us to revise our views.

When in the course of philosophical debate each side assumes the burden of proof to be on the other—as evidently is the case with the stereotypical nominalist and the stereotypical anti-nominalist as we have described them—the debate often tends towards stalemate. As will be seen below, such has been the tendency in three rounds of debate over arguments for nominalism. (For more on the issue of burden of proof, see Rosen 1992: chapter 2.)

b. *The Epistemological Argument: Original Version*

The most straightforward version of the basic epistemological argument for nominalism would simply attempt to give an explicit formulation and defence of the major premiss of the syllogism, of a causal theory of knowledge. If one looks for such an explicit formulation and defence of such a causal theory, one will not find it in later works by avowed nominalists, but will be led back by citations in those works to Benacerraf's original paper. One will not find it there, either, but will be led back in turn to some of the specialist literature on epistemology or theory of knowledge that it cites, notably the work of Alvin Goldman, beginning with Goldman (1967).

While in very many cases causal connections between the knowing subject and known object are obviously *present*, the specialist work like Goldman's provides examples of cases where causal connections seem absolutely *crucial*. Work like Goldman's then attempts to extract from such examples and formulate explicitly a causal theory of knowledge detailing just what kind of causal connections are crucial and in just what way they are crucial. So superficially, at least, it appears that this specialist work provides just what the nominalist is looking for.

But all this is only a superficial appearance. In reality, no causal theory of knowledge of a kind exploitable for nominalistic purposes receives any significant support in the specialist literature. The reasons why may be summed up under four heads, in order of increasing importance: (i) the status of Goldman's theory; (ii) the scope of the particular problem to which that theory was supposed to provide a solution; (iii) the nature of the particular problem to which that theory was supposed to provide a solution; (iv) the nature of the general problem addressed in the specialist literature, of which that particular problem was but one aspect.

As to point (i), Benacerraf's work was fairly closely tied to the specialist literature of the time, in which Goldman's theory was considered a promising candidate for a solution to the problems it addressed. Subsequent nominalist appropriations of Benacerraf's work have tended to float free of any mooring in the specialist literature. Had the development of that literature been followed more closely, it would have been seen that Goldman's theory, however promising it seemed initially, increasingly encountered difficulties. It has by now long since come to be considered less satisfactory than several rival, non-causal theories. (In this connection see Maddy (1984*a*).)

Points (ii) and (iii) require some background. Philosophers since antiquity had held that there is a distinction to be made between genuine knowledge and mere right opinion or true belief. For a belief to rank as knowledge, it must not only be true, but must also be appropriately warranted or justified. A famous note of Edmund Gettier, Gettier (1963), provided counter-examples showing that the three necessary conditions, truth and belief and justification, are not by themselves sufficient for knowledge, and that there is a need for a fourth condition. A typical Gettier example might go as follows.

Zack enters a room, looks towards a table at one end of it, sees an apple, and forms the belief that there is an apple on that table. This belief is justified if anyone's belief that there is something of a certain sort in a certain place, formed on looking towards that place and seeing something of that sort, is ever justified. The belief is also true, for there is an apple on the table. However, Zack is being tricked by Yolanda, who wishes to prove a philosophical point. The apple Zack sees is not the apple on the table, for Yolanda has placed a mirror on the table in front of *that* apple, positioning it in such a way that anyone entering the room and looking towards the table will see a reflection of *another* apple on a matching table at the other end of the room.

Gettier would submit that Zack's justified true belief is not knowledge, and would ask what fourth condition is missing. This is the problem Goldman was addressing. His suggestion was that what is missing in this case is an appropriate causal relation between its being true that there is an apple on the table and Zack's believing that there is an apple on the table. That is his candidate solution.

Returning now to point (ii), it may be noted that the Gettier examples pertained to cases of empirical knowledge of contingent facts about concrete entities, and that Goldman's suggestion was addressed only to such cases, so that it was silent about precisely the cases of most interest in connection with the issue of nominalism.

And turning to point (iii), it may be noted that the problem Goldman was addressing pertained to what is required in order for a justified true belief to rank as knowledge, and that he proposed causal connections as part of an answer to that question, and not to any question about what is required in order for a true belief to rank as justified. This means that, confronted by a nominalist with Goldman's theory, an anti-nominalist could cheerfully say: 'Very well, then, let's compromise. Concede that your disbelief in numbers is unjustified, and that my belief in them is justified, and I'll concede that my justified belief in numbers technically speaking can't be called "knowledge" in the strictest sense of the term. I can live with that if you can.' Plainly 'compromise' on these terms would be tantamount to surrender by the nominalist side. But nothing in Goldman's original paper allows the nominalist to demand any better terms than these.

Turning finally to point (iv), the general problem addressed in the specialist literature, of which the Gettier problem is merely one particular aspect, is that of conceptual analysis of the ordinary notion of knowledge, of the meaning of 'to know'. Generally speaking, if a candidate analysis of the meaning of a term has the implication that many or most users of a term are rather obviously misusing it much or most of the time, then it cannot be considered a very promising candidate analysis. But an unrestricted causal analysis of 'know' would imply that every time an elementary or secondary school teacher says something like, 'Xavier knows his multiplication facts', or 'Wilhelmina knows a good deal of algebra', the teacher is making a rather obviously false claim about causal connections between the students and numbers. The very features that would make an unrestricted causal analysis exploitable for nominalistic purposes would virtually exclude it from serious consideration as a candidate conceptual analysis.

In any case, as the discussion two paragraphs back shows, if there is to be a causal theory of knowledge exploitable for nominalistic purposes, it will have to be a causal theory of *justification* developed independently of, and to a significant extent in opposition to, the specialist literature. It is no easy task to formulate (let alone defend) a plausible candidate for such a theory. Perhaps the most important difficulties are the following two. On the one hand, there is the problem of what, for a given proposition P, must be the status of the higher-order proposition:

(*P**)   there is an appropriate causal connection between its being true that P and its being believed that P

in order for the belief that P to be a justified belief that P. On the other hand, there is the problem of what kinds of causal connections between its being true that P and its being believed that P count as 'appropriate'.

Taking the first problem first, surely it can't be required that $P*$ must be true in order for the belief that P to be justified, since this would imply that no belief can be justified if it is false. For $P*$ can't be true if P is false: there can't be an appropriate causal relation—or for that matter, an inappropriate causal relation or a non-causal relation—between its being believed that P and its being true that P, if it isn't true that P. But surely some false beliefs are justified. For instance, even if there hadn't been an apple on the table, Zack's belief that there was would still have been justified, given what he saw when he looked towards the table.

Nor can it be required that $P*$ must be justifiably believed in order for the belief that P to be justified. For such a requirement would involve an infinite regress: for P to be justifiably believed, $P*$ would have to be justifiably believed, for which $P**$ would have to be justifiably believed, and so on *ad infinitum*. But surely, for instance, Zack can justifiably believe there is an apple on the table without engaging in zillionth-order reflection on the causes of this belief about the causes of his belief about the causes of his belief about . . . the apple. However, it would seem very odd to require merely that $P*$ must be believed, thus holding that an unjustified belief that $P*$ could be what made the belief that P justified. It would also seem very odd to require merely that $P*$ must not be disbelieved, thus allowing that omission to consider the question whether it is the case that $P*$ could be what made the belief that P justified.

There remains the possibility that what is required for a person's belief that P to be justified is for that person to be *justified in believing $P*$*, in a sense in which it is possible to be justified in believing something without actually believing it. (Holmes may be justified by the totality of the evidence in believing that Watson is the culprit even if he recoils from the thought and fails to embrace this conclusion.) But even this seems an implausible requirement.

To get a sense of the problem, consider a justified belief that has been held by people of all ages and nations: if one sees a full moon in the sky one night, there will be another full moon to be seen there about twenty-eight or twenty-nine days later, give or take a day or two. This purely empirical belief has been accompanied by the most varied and most bizarre beliefs about celestial mechanics (about the causes of visible phenomena in the night sky), and about cognitive psychology (about causes of human beliefs). Even confining attention to Europe, and to the seventeenth and eighteenth centuries, one finds in circulation no end of astronomical theories: Aristotelian, Ptolemaic, Copernican, Tychonian, Keplerite, Newtonian. Adherents of some of these emphasized that they 'feigned no hypotheses' about underlying mechanisms and sought only to 'save the phenomena'. Adherents of others posited crystalline spheres or other apparatus that seem as weird to us as the Einsteinian theory of warped space would doubtless have seemed to them. And if beliefs about celestial mechanics were odd, how much odder were beliefs about cognitive psychology. Consider, for instance, the philosophical teachings of Leibniz, Malebranche, Berkeley, and others, whose pre-established harmony, occasionalism, immaterialism, and related doctrines all in one way or another denied that material heavenly bodies were the real causes of people's perceptions when they turn their eyes skywards. Or consider one influential school, the Cartesian, which held that much of belief was a matter of free will, and thus had no external causes and in a sense no causes at all. It would seem wrong to claim that holding quaint and curious additional beliefs about the causes of the moon's phases and of our expectations about them somehow undid the simple empirical justification for the simple empirical belief about the night sky. It would seem even more wrong to claim that holding such quaint and curious beliefs *constituted* that justification.

Let us suppose that this first problem in formulating a causal theory of justification can somehow be overcome, and turn to the second problem, that of 'appropriateness'. This also is not easy to define in a way that is neither too strong nor too weak. On the one hand, it cannot be held to be sufficient that the fact that P should be one that would be cited in the causal explanation of the formation of the belief that P. Or at least, a theory on which this is a sufficient condition for justification would be too weak for nominalistic purposes. To understand why, suppose a child counts out three pebbles, then counts out two more, then counts them all out together, and as a result forms the belief that three plus two is five. In the explanation of how the child came to believe this, the arithmetical fact that three plus two is indeed five will be cited in explaining why the child got five on the final count, along with the mineralogical fact that pebbles don't coalesce or evaporate or the like between successive counts only minutes apart, and various other facts. In a larger sense, an explanation of the formation of almost any belief that really gets down into physiological detail is likely to have to cite some fairly sophisticated mathematics, as was noted long ago in Steiner (1975).

On the other hand, it cannot be taken to be a necessary condition for appropriateness that the fact that P should help cause the belief that P. Or at least, such a requirement would be too strong for the purposes of a nominalist who does not want to go on from nominalism to a more general scepticism. For the requirement would make all beliefs about the future unjustified. Similarly, to take it to be a necessary condition for appropriateness that the entities involved in the fact that *P* help cause the belief that *P* would still seem too strong, since it would rule out the possibility of justified belief in future entities.

Even to take it to be a necessary condition for appropriateness that some entities of the same sort as those involved in the fact that *P* help cause the belief that *P* would arguably be too strong. For suppose a cosmological theory implies that in certain conditions that will only be realized just prior to the final collapse of the universe, and in particular will only be realized after the extinction of the human species, certain particles—call them *eschatons*—will be produced that will differ in fundamental ways from everything else that has existed previously. If the theory is sufficiently well confirmed, we might thereby come to have reason to believe that eschatons will exist (or do exist in our future), even though according to the theory they are quite unlike and much stranger than any sort of particle we have encountered or interacted with so far, or ever will encounter or interact with.

More generally, anti-nominalists of all stripes can be expected to point out that, in the absence of a detailed formulation of a causal theory of justification, one may wonder whether such a theory really can be formulated so as to have just the consequences a nominalist would want, and not further, undesired, sceptical consequences. Beyond this, different kinds of anti-nominalists—those who hold that we do after all in a sense have causal connections with some sorts of abstracta, those who assimilate mathematical and ontological knowledge to knowledge of logic and definitions, and more stereotypical anti-nominalists as we have depicted them—will have different kinds of concerns.

The stereotypical anti-nominalist is likely to find the emphasis on problems of details of formulation in the above discussion a bit misplaced. For supposing those problems to be overcome, there would still remain what to stereotypical anti-nominalists, who assume the burden of proof to be on the nominalist side, must seem the greatest problem, namely, the problem of explaining why we are supposed to have more confidence in the causal theory of justification than in established mathematics, or such commonplace judgements about it as the judgement, 'That π is transcendental has been known since the nineteenth century.'

Emphasis on details of formulation also tends to seem misplaced from a stereotypical nominalist perspective. Nominalists are not unaware of some, at least, of the unresolved problems we have been reviewing.

(Benacerraf himself already commended to his readers' attention the work of his student Mark Steiner, cited above.) But though not unaware, they seem typically to have been unimpressed. The review, Hart (1977), of Steiner's book seems atypical only in the vehemence of its rhetoric:

[I]t is a crime against the intellect to try to mask the problem of naturalizing the epistemology of mathematics with philosophical razzle-dazzle. Superficial worries about the intellectual hygiene of causal theories of knowledge are irrelevant to and misleading from this problem, for the problem is not so much about causality as about the very possibility of natural knowledge of abstract objects.

Like W. D. Hart, stereotypical nominalists do not take the burden to be on the nominalist party to produce a detailed theory of knowledge or justification: rather, they take it be on the anti-nominalist party, to explain in detail how anything we do and say on our side of the great wall separating the cosmos of concreta from the heaven of abstracta can provide us with knowledge of the other side. Thus regardless of the outcome of attempts at a detailed formulation of a causal theory of justification, a stalemate, resulting from opposite assumptions about where the burden of proof lies, is to be expected. (For more on this topic, see Rosen 1992: chapter 3.)

c. The Epistemological Argument: Refined Version

An important attempt to recast the nominalist's general epistemological challenge in a way that will avoid becoming bogged down in questions of detail about the formulation of particular theories of knowledge has been made by Hartry Field (in §2 of Field (1988), reprinted as chapter 7 of Field (1989), and in §4.A of chapter 1 of Field (1989)). His challenge, as he says, does not involve 'the term of art "know"'. Rather, it is supposed to be a challenge to our ability to *explain the reliability* of our mathematical beliefs, to explain the correlation between what we believe about mathematicalia and what is true about them.

Field's exposition of the challenge begins with the observation that, from an anti-nominalist standpoint, the following holds (as a general rule, with some exceptions):

(i)   when mathematicians believe a claim about mathematicalia, then that claim is true

This may be called the 'reliability thesis'.

But now look what it says: there is a pervasive correlation between two quite different realms of facts, namely, between certain facts about human beliefs, and certain facts about abstract entities. Surely it is not reasonable to accept such a correlation as a brute, inexplicable fact: there may be brute facts about abstract entities; there may even be brute facts about human beliefs; but the correlation between two such distinct realms of facts cannot just be accepted as a brute fact not to be explained in terms of anything more basic. Field concludes:

(ii)   if the reliability thesis is true, then it must be explained

(Though Field maintains that his challenge is 'not to our ability to *justify* our mathematical beliefs', and though he does not explicitly rely on any causal theory of justification, the implicit suggestion in (ii) can hardly be anything but this, that if the reliability thesis cannot be explained, then continued belief in claims about mathematicalia is unjustified.)

But now here's the problem: the causal inactivity and impassivity of abstracta seems to rule out the possibility of any sort of explanation of the reliability thesis. The facts about the abstracta can't be the causes of the facts about the beliefs, since abstracta are inactive. The facts about the beliefs can't be the causes of the facts about the abstracta, since abstracta are impassive. For the same reason, the two sets of facts can't be effects of a common set of causes. A non-causal explanation of the reliability of logical and analytic knowledge about abstracta (or for that matter, about anything) may be possible, since such knowledge is ultimately knowledge of language. But what could a non-causal explanation of the reliability of synthetic beliefs about abstracta conceivably be like? Field does not claim to have established conclusively that the reliability of beliefs about abstracta is in principle inexplicable, but he does express deep pessimism that:

(iii)   the reliability thesis cannot be explained

How is the anti-nominalist to respond? Anti-nominalists of all stripes can be expected to point out that, in the absence of a detailed explanation of the reliability of human beliefs about concreta, one may wonder whether the challenge will have just the force a nominalist would want, and not a further, undesired, sceptical force. Beyond this, different kinds of anti-nominalists will make different kinds of responses. The challenge is in fact not directed against a position like Maddy's, which Field has addressed separately elsewhere (in Field 1990); nor is it directed against a position like that of the St Andrews School, which again Field has addressed separately elsewhere (beginning with Field (1984*b*), reprinted as chapter 5 in Field (1989)). It is appropriate, therefore, to consider it from the perspective of the stereotypical anti-nominalist position as we have described it, or from that of an anti-nominalist position like Quine's. From such a perspective, the chief reservation will be over the concept of justification on which Field's challenge seems tacitly to rely; but there will be further questions about the notion of truth involved, and these we take up first.

The reason some discussion of the notion of truth seems unavoidable is that truth is mentioned in the reliability thesis, and is a philosophically controversial notion—with Field (like Quine) being among the main controversialists, especially in two major and difficult papers (Field 1972, 1986). Field accompanies his presentation of his challenge with some prefatory remarks on truth, partially in response to critical comments of William Tait. The main issue is a simple one. In the second of his papers on truth, Field expressed some sympathy for the view that a more robust notion of truth than the disquotational may be needed for some purposes. But what Field insists in response to Tait (1986) is that for purposes of formulating his challenge, he needs only a certain ability to state generalizations that is provided by the disquotational notion of truth: he needs no 'heavy duty' notion of truth, as he puts it. The roughest and briefest of expositions of the disquotational theory of truth—which in any case will be needed for other purposes later—will suffice to indicate what 'ability to state generalizations' is involved.

The word 'true', taken in the disquotational sense, is used not for reporting substantive relations of correspondence between language and extra-linguistic reality, but for the intra-linguistic purpose of undoing quotation, the operative rule of language being as follows:

(iv)   '_____'   is true   if and only if   _____

of which the canonical example is:

        'snow is white'   is true   if and only if   snow is white

Thus to assert that '_____' is true in the disquotational sense is not to add anything substantive to the bare assertion that _____: the notion of truth in the disquotational sense is redundant in application to a single explicitly quoted sentence. What saves it from being wholly redundant and so wholly useless, is mainly that it provides an ability to state generalizations like:

(v)   everything Fermat believes is true

Whatever is put in the blank '_____', (v) implies:

(vi)   if Fermat believes '_____', then '_____' is true

And (iv) and (vi) together imply:

(vii)   if Fermat believes '_____', then _____

Thus (v) expresses in a single assertion the indefinitely long list of assertions:

> if Fermat believes, 'there are $x, y, z > 0$ such that $x^3 + y^3 = z^3$', then there are $x, y, z > 0$ such that $x^3 + y^3 = z^3$

> if Fermat believes, 'there are no $x, y, z > 0$ such that $x^3 + y^3 = z^3$', then there are no $x, y, z > 0$ such that $x^3 + y^3 = z^3$

> if Fermat believes, 'there are $x, y, z > 0$ such that $x^4 + y^4 = z^4$', then there are $x, y, z > 0$ such that $x^4 + y^4 = z^4$

> if Fermat believes, 'there are no $x, y, z > 0$ such that $x^4 + y^4 = z^4$', then there are no $x, y, z > 0$ such that $x^4 + y^4 = z^4$

and so on.

After this short digression we may return to the question how an anti-nominalist (with a position in the general neighbourhood we indicated) is to respond to Field's challenge. The triad (i), (ii), (iii) is inconsistent. For an anti-nominalist, giving up (i) would simply be giving up. Total rejection of (ii) seems quite unattractive, but refutation of (iii) by providing a total explanation seems quite unfeasible. Presumably what the anti-nominalist will have to try to do will be to give some combination of partial explanation with reasons for rejecting demands for further explanation (that is, reason for holding that the failure to meet the demand for further explanation does not render beliefs unjustifiable).

One demand that could be rejected straight off would be a demand that the anti-nominalist give a uniform explanation of human reliability across all the diverse areas of enquiry that involve abstracta, from pure mathematics to theoretical physics to political economy to literary criticism. Such a demand is unreasonable because on any view the processes of belief formation are not uniform across diverse areas of enquiry: different areas use different methods, reliable in different degrees and for different reasons. If the anti-nominalist can explain the reliability of our beliefs about one sort of abstracta in one particular area, then the challenge is met, as regards that sort of abstracta in that particular area. Field himself makes implicit partial acknowledgement of this fact by stating the reliability thesis specifically for mathematicians' beliefs about mathematicalia; sometimes he states it even more specifically for (pure) set-theorists' beliefs about (pure) sets, and we may limit our attention to that test case.

Field himself then suggests a considerable further reduction of the problem: he does not question the explicability of the reliability of judgements about what follows logically or analytically from what. Thus whatever area is considered, the issue is not really about the reliability of arbitrary beliefs within that area, but rather about axiomatic beliefs, ones that are not believed simply because they follow logically or analytically from other, more basic beliefs. The challenge is thus to explain the reliability of axiomatic beliefs in set theory, to explain the correlation between axioms of standard set theory being believed and their being true.

But just how general and pervasive a phenomenon is involved here? On any elegant formulation of the axioms of standard set theory, they are very few. Those few can quite naturally be regarded as partial definitions or descriptions of a certain mathematical structure, of a universe of sets arranged in a certain way, of what is often called the 'full cumulative hierarchy' of sets. When the axioms of standard set theory are so regarded, then there is in effect just one, sole, single, unique axiom:

(viii)   the full cumulative hierarchy of sets exist

Now one can hardly speak of a 'pervasive' correlation between the items on two lists when there are only a very few items on each; and when there is only one item on each, one cannot speak of a 'correlation' at all. What we have before us is not a correlation, but just a conjunction. The first conjunct is:

(ix)   it is true that the full cumulative hierarchy of sets exist

The second conjunct is:

(x)   it is believed that the full cumulative hierarchy of sets exist

The challenge reduces to the demand for an explanation of this conjunction.

We have already noted (in article 1.b) that, assuming that mathematicalia of a certain kind exist, it makes very questionable sense to demand why they do, as if they could easily have failed to do so (just as, assuming they do not exist, it makes very questionable sense to demand why they don't, as if they could easily have done so). The demand for an explanation of conjunct (ix) can therefore reasonably be rejected. Moreover, at least a very good beginning towards an explanation of conjunct (x), of how and why standard set theory came to be believed, is given in standard histories.

Is there then anything left that needs explaining but hasn't been explained? Well, the *connection* between the two conjuncts has not been explained. Without such an explanation it may appear mere accident or luck that the theory we have come to believe is a theory that is true. The implicit suggestion will then be that if this has to be acknowledged to be just a lucky accident, then continued belief in standard set theory is not justifiable.

How is the anti-nominalist to respond? One can hardly avoid acknowledging that standard set theory is the end product of an immensely complex historical process that could have gone differently in countless ways. It was lucky that Cantor came along when he did with the key concepts; that opposing forces, which kept him from obtaining a major professorship and from publishing in some major journals, did not silence him altogether; that unlike some of his forerunners, he found contemporaries with the capacity to understand and appreciate his theories. But what we have just said about the cumulative hierarchy of set theory, in which Field does not believe, could equally be said about the warped space of general relativity, in which he does believe. Surely it is to a large degree a matter of luck that Einstein came along when he did with the key concepts; that the Nazi campaign against 'Jewish physics', with Einstein as its foremost target, did not succeed; that the remark, attributed by legend to Einstein himself, that only a dozen people in the world would have the capacity to understand the mind-bending implications of warped space, proved unfounded. If there is an argument for anything in the fact that accident or luck plays a large role in the history of science, it is an argument not just against set theory but against general relativity as well: it is an argument not for nominalism in particular, but for scepticism in general.

But is there a cogent argument here? On the one hand, it must be insisted that it is simply to a large extent a matter of accident that human beings with the capacities they have ever existed, and a matter of luck that some of them enjoyed favourable conditions for the exercise of those capacities: people might easily never have been lucky enough to enjoy the leisure and the freedom to develop (what Quine calls) the 'refinement of common sense which is science', with its refinement of commonsense standards for the choice of theories; or even having done so, people might easily never have been so lucky as to hit upon the key concepts needed to formulate certain theories and place them among the candidates from which to choose. The demand for a total explanation, showing that all this was inevitable and foreordained, must be rejected; but so must any suggestion that the lack of such an explanation makes our beliefs unjustifiable, whether in set theory or in general relativity. On the other hand, standard histories go a very long way towards explaining why, given human capacities and given favourable conditions for their exercise, we came to believe as we have, whether in set theory or in general relativity.

So once again we ask: is there anything left that needs explaining but hasn't been explained? And once again we answer: well, there is a *connection* that has not been explained. It is the connection between set theory's being something that creatures with intellectual capacities and histories like ours might, given favourable conditions for the further exercise of their capacities, come to believe, and set theory's being something that is true. Standard set theory, once it has been thought of, may be a very good theory by scientific standards. But those standards, on most accounts (including Quine's), include simplicity. And that can only mean simplicity as felt by creatures with capacities and histories like ours. But one may then demand an explanation:

[W]hy should one believe that the universe of sets . . . is so nicely arranged that there is a preestablished harmony between our feelings of simplicity, etc., and truth?

The implicit suggestion is that in the absence of a response, continued belief in the truth of standard set theory would be unjustified. What is being asked is thus in effect:

(xi)   Granted that belief in standard set theory
            is justified by scientific standards,
        is belief in the truth of standard set theory justified?

By the time we arrive at this formulation of the problem, we have left Field's formulation far behind: the above-quoted formulation is in fact Benacerraf's (writing with Putnam in Benacerraf and Putnam (1983: editorial introduction, §9), and as always describing, not advocating, a nominalist position). Benacerraf makes in connection with his formulation three highly suggestive passing remarks.

Benacerraf's first remark hints that the problem may not be peculiar to mathematics: 'It is hard enough to believe that the natural world is so nicely arranged that what is simplest, etc. by our lights is always the same as what is true (or, at least, is generally the same as what is true) . . .' And indeed, there are contemporary philosophers who find this not just hard but impossible to believe. Just as the nominalist asks why the fact that we can most efficiently cope with physical systems by representing them to ourselves as approximate realizations of ideal mathematical systems should be considered a reason for believing that in truth there are such systems, so a neo-Machian could ask why the fact that we can only effectively cope with what we perceive by representing it to ourselves as part of a system also having parts we do not perceive should be considered a reason for believing that in truth there are such systems. Thus again the argument for nominalism threatens to collapse into an argument for scepticism. This first remark suggests that one should put the challenge (xi) in a more general form:

(xii)   Granted that belief in some theory
                 is justified by scientific standards,
            is belief in the truth of that theory justified?

Benacerraf's second remark hints that the problem will not seem serious to those who do not adopt a certain conception of truth he calls 'realist': 'The very question . . . presupposes . . . a realist notion of truth.' This remark reminds us that we are supposed to be considering the question (xii) from a standpoint that admits and relies on only a disquotational notion of truth. With this understanding, the question becomes:

(xiii)   Granted that belief in some theory
                 is justified by scientific standards,
            is belief in that theory justified?

Once it is put in this last form, it becomes clear that the question or challenge presupposes a 'heavy duty' notion of 'justification'—one not just constituted by ordinary commonsense standards of justification and their scientific refinements—of the kind the stereotypical anti-nominalist regards with suspicion. To put the matter another way, once it is put in this last form, it becomes clear that the question or challenge is essentially just a demand for a philosophical 'foundation' for common sense and science—one that would show it to be something more than just a convenient way for creatures with capacities like ours to organize their experience—of the kind that Quine's naturalized epistemology rejects. And this explains a third remark of Benacerraf's: 'Quine . . . tends to pooh-pooh this sort of question.'

If Field's challenge does ultimately reduce to the form Benacerraf considers, then it presents a stereotypical anti-nominalist position, or a position in the general vicinity of Quine's, with no obvious threat of internal collapse. It does, however, draw attention to just what and just how much such a position is renouncing when it renounces any ambition to provide common sense and science with a 'justification' by some exterior, superior, ulterior standards, when it renounces the ambition to provide a philosophical 'foundation' for common sense and science. In the light of the challenge, such an anti-nominalist position is likely to seem adequate or inadequate according as one takes the burden of proof to be on the side of its opponents or of its proponents. And thus stalemate threatens again.

d. *The Epistemological Argument: Semantic Analogue*

Today nominalist arguments from the theory of knowledge are usually mentioned in the same breath as arguments from the theory of reference. There are hints of such an argument already in Benacerraf (1973), and explicit such arguments appear in the literature soon thereafter, as in Jubien (1977). Debate over reference will be treated only briefly here, since in so many ways it is just a replay of debate over knowledge. To begin with, the basic argument is again a syllogism or enthymeme. The explicit or tacit major premiss is a causal theory of reference. The minor premiss is the causal inactivity of abstracta. The conclusion is again one that obliquely undermines belief in abstracta without directly contradicting it: the conclusion is that we can make no reference to abstracta.

The most straightforward argument for the major premiss would cite the specialist literature for examples. While in very many cases causal connections between a speaker or writer and the objects to which he or she makes reference are obviously present, the specialist literature provides examples of puzzle cases where causal connections seem absolutely crucial. They seem crucial when speakers succeed in everyday contexts in referring to a certain object using a descriptive phrase that is largely false of it, as when someone succeeds in referring to the man over there drinking ginger ale, though describing him as 'the man over there drinking champagne'. They seem crucial when writers succeed in scientific contexts in referring to objects of certain kind using a descriptive theory that is largely false of them, as when nineteenth-century chemists succeeded in referring to atoms, though theorizing that atoms are indivisible. They seem crucial whenever speakers and writers succeed in referring to objects using proper names, which J. S. Mill long ago claimed and Saul Kripke has persuasively argued are not equivalent to descriptions of any kind, true or false. And thus it is that in the work of Kripke and others throughout the specialist literature one again and again encounters what commentators often call 'causal theories of reference'.

However, on closer examination this specialist work proves to provide no genuine support for nominalism. (One of Kripke's examples actually concerns an abstract entity, a Lie group named 'Nancy', discussed in Kripke (1972: 116 n.), with no suggestion that the views on reference he advocates in any way make such a case problematic.) Thus the solution offered in Kripke (1972) to the problem how a category of names that are not simply abbreviations for descriptions could work is, roughly and briefly, that on the first occasion the name would be used with the intention of referring to whatever a certain description refers to, while on each subsequent occasion the name would be used with the intention of referring to whatever was referred to on the latest previous occasion, so that the initial description would not be permanently associated with the name, and might soon be forgotten. This theory may perhaps be said to posit a causal chain connecting later and later users of a name with earlier and earlier ones, back to the initial coiner of the name (though Kripke's account is entirely silent on the question of whether the human choices involved are products of causal determinism, random happenstance, or free will of a kind that is neither the one nor the other of these); but it says nothing about any causal chain leading back from the initial coiner of the name to the entity named: it is not required that the initial description be couched in terms of or be accompanied by causal connections with the bearer of the name. Indeed, neither in Kripke's nor in any other important discussion in the specialist literature is it denied that one can usually succeed in referring to an object simply by offering a true description of it: the 'causal theories of reference' in the specialist literature are addressed to the question how one can sometimes succeed in referring to an object without offering a true description of it.

Indeed, since the discussions in the specialist literature were intended as contributions to the analysis of ordinary judgements of when reference is successfully achieved, and not as exposés of massive error in such judgements, such specialist work never should have been expected to provide support for nominalism. Anyone who wishes to formulate a causal theory of reference suitable for nominalistic purposes must do so largely independently of the specialist literature, if not in opposition to it. The task of formulating such a theory is no easy one. On the one hand, it cannot be said that a term refers to whatever prompts its use, since not only would this make every term multiply ambiguous (since every use of a term has *many* causes), but it would also make it impossible to misidentify any object directly present to the senses. On the other hand, it cannot be claimed that in every case the referent of a term must be among the causes of its use, since this would make reference to future entities impossible, and would result in a theory of reference with implications going well beyond nominalism in the direction of a more general scepticism.

If asked to respond to the challenge of the causal theory of reference, the stereotypical anti-nominalist will say that one cannot respond properly until the theory in question has been properly developed and presented. Above all, the stereotypical anti-nominalist will say that one cannot respond properly until it has been explained why one should have more confidence in such a theory of reference than in such ordinary judgements about standard mathematics as, 'The symbol "π" has been used to refer to the number pi since the 18th century'—to say nothing of standard mathematics itself.

The stereotypical nominalist, by contrast, takes it to be obvious that there is a serious problem about how anything anyone says or does can establish a relationship of reference between words on our side of the great wall dividing the concrete from the abstract and entities on the other side. Consider the symbol 'o' or the word 'zero', for instance. Surely it didn't always refer to the number zero, and didn't have to refer to the number zero. It could have referred to the number one instead. It could have referred to the philosopher Zeno or the emperor Nero. It could have had a linguistic function other than referring, perhaps as an interjection like 'oh'. It could have had no linguistic function at all. 'But what', the nominalist will ask, 'can one do or say over here in this world to make it refer to something over there in the other world, and to one such thing rather than another?' (It may be noted in passing that it should be some small embarrassment to nominalists who press the challenge in these terms that their questions are framed as questions about the reference of linguistic *types*, entities in which they supposedly do not believe; but we will not stop to press the point.)

One can, of course, offer a description: 'Zero is the least natural number.' This fixes reference of 'zero', provided the meanings of 'natural number' and 'less than' are fixed. But the stereotypical nominalist takes it to be obvious that there is a serious problem about how any usage, anything one says or does on our side of the great wall, can fix the meaning of a predicate supposed to apply to entities on the other side, can make one interpretation of it uniquely correct, and any and all others erroneous. (The issue thus becomes one of 'semantics' not only in the broader sense in which that subject includes the theory of reference, but also in the narrow sense in which it includes only the theory of meaning.)

The most obvious constraint our usage places on how we are interpreted is that we would expect the correct interpretation to be one on which our assertions generally come out meaning something true. But works by sympathizers with nominalism, notable among them Hodes (1984*b*), have cited numerous examples as showing that everything we say about the number zero or the whole system of natural numbers can be reinterpreted in such a way as to make it true, while making 'less than', for instance, mean something other than less than, and making 'zero', for instance, refer to something other than zero. The simplest example of this problem of multiple interpretations makes use of the permutation function:

$$\pi_{a,b}(c) = b, \text{ if } c = a$$
$$= a, \text{ if } c = b$$
$$= c, \text{ otherwise}$$

One can then take the usual numeral for each number n to refer not to n itself but to $\pi_{0,1}(n)$, so that in particular 'zero' is taken to refer to one. More precisely, one can do this provided one makes compensatory changes in the interpretation of various predicates. For instance, 'm is odd' and 'n is even' would be interpreted as:

$$\pi_{0,1}(m) \text{ is odd}$$
$$\pi_{0,1}(n) \text{ is even}$$

which amount to:

    m is odd and distinct from one, or is zero
    n is even and distinct from zero, or is one

The example can be made universal, changing the reference of all terms. If negative as well as positive integers are admitted, then it is especially easy to describe such a universal example: it suffices to use in place of $\pi_{0,1}$ a different permutation function, $\psi(n) = n + 1$.

Drawing on more and more technical results from mathematical logic and foundations, a series of more and more dramatic versions of the basic example can be given, and are given in literature (though in the best-known discussions there, the morals drawn are not quite nominalistic ones). As some of these examples presuppose more knowledge of technicalities than we wish to assume, we will mention them only briefly. First, the referents of all numerals can be taken to be entities other than natural numbers, and hence the extensions of 'is a natural number' and 'is less than' and so forth can be taken to be disjoint from their usual extensions. For (as already mentioned in article 1.b), set-theoretic surrogates are available not only for natural numbers, but for virtually all mathematicalia. Indeed, in the cases of natural numbers, real numbers, and many others, many different systems of set-theoretic surrogates are available (of which the two most important will be briefly indicated in article B.1.a). This situation has been much discussed ever since Benacerraf (1965).

Second, the predicates 'is a natural number' and 'is less than' and so forth can be so interpreted that, though everything that is usually said about natural numbers and the order relation on them remains true, the entities interpreted as 'natural numbers' are not ordered in the usual way by the relation interpreted as 'less than'. Indeed, the interpretation can be so chosen that the entities interpreted as 'natural numbers' are more numerous than the usual natural numbers; and inversely, an interpretation can be so chosen that the entities interpreted as 'real numbers' are less numerous than the usual real numbers. These last results involve an important theorem of mathematical logic (discussed briefly in article B.4.a). Such results have been much discussed since Putnam (1980).

A less technical, but in the opinion of many philosophers deeper, problem is presented in Kripke (1982). Readers are warned in that book that the argument there outlined should not be attributed without qualification either to the author, Saul Kripke, or to his subject, Ludwig Wittgenstein. In consequence of this warning it has become customary to call it 'Kripkenstein's argument'. Kripkenstein's problem is that no finite list of examples of usage can ever determine that one means sum by 'sum', rather than something that differs from the standard sum for larger numbers than one has so far considered, perhaps by adding 1 per cent to sums over $10^{10^{10}}$. Verbal definitions are of no help, since the predicates in them are likewise subject to distorted interpretation, as the cited book demonstrates at some length. Nor does appeal to people's behavioural dispositions help, since people's behavioural disposition is in fact to give the wrong answer (or no answer at all) more and more often as the summands get larger and larger, and in any case the problem of interpretation or meaning is one of determining what makes one answer rather than another not just the one people do give but the one they ought to give, if they are to keep faith with their previous understanding of the relevant vocabulary. Where behavioural dispositions themselves will not help, neither will neural causes underlying them, or introspectable images accompanying them. Or so Kripkenstein argues.

Now the first thing the anti-nominalist will want to say in response to all this is that reinterpretations preserving truth but altering reference are possible in the case of concrete entities, too. Indeed, the definition of $\pi_{a,b}$ above is perfectly general, and in no way depends on a,b being natural numbers. Thus one could reinterpret our language so that 'Adam' refers to Eve and 'Eve' to Adam, provided the obvious compensatory changes in the interpretations of predicates are made. For instance, 'y is male' and 'x is female' would be interpreted as:

$$\pi_{\text{Adam,Eve}}(y) \text{ is male}$$
$$\pi_{\text{Adam,Eve}}(x) \text{ is female}$$

which amount to:

y is male and distinct from Adam, or is Eve
x is female and distinct from Eve, or is Adam

If conglomerates are admitted, then it is especially easy to describe an interpretation making the example universal, changing the reference of all terms. It suffices to use in place of $\pi_{0,1}$ a different permutation function, $\psi(a) =$ the contents of the region of space complementary to that occupied by a.

Beyond this, there are further examples, of which a few may be cited here. For one instance, our usage of 'the Rock of Gibraltar' does not seem to be so precise as to determine the exact boundaries of its referent on a scale of centimetres. Nothing seems to settle just how far below surface level it extends, or whether outcroppings along its edges that are covered at high tide and exposed at low tide are part of it. To put the matter another way, there seem to be any number of regions with exact boundaries whose contents could be taken as the referent of 'the Rock of Gibraltar', compatibly with all our usage of that term. The situation is at least roughly analogous to that in Benacerraf's example, where there are any number of systems of set-theoretic entities, any one of which could be taken to be the referents of 'zero, one, two, . . .', compatibly with all our usage of those terms.

For another instance, while Kripkenstein's problem was presented for a predicate, '. . . is the sum of . . . and . . .', applying to abstracta, it was intended all along to apply also to predicates like '. . . is green' applying to concreta: no finite list of examples of usage can ever show that one means green by 'green', rather than something that differs from standard greenness at times far in the future, say by taking in bluer and bluer entities. All the considerations that applied to 'sum' apply equally to 'green'. Moreover, the example can be adapted from the satisfaction of predicates to the reference of terms. No finite list of examples of usage can ever show that one refers to the Rock of Gibraltar by 'the Rock of Gibraltar', rather than something whose past and present temporal stages coincide with those of the Rock of Gibraltar, but whose later ones diverge from it more and more, perhaps drifting further and further southward.

For yet another instance, perhaps the oldest and certainly the most extensively debated, there is the claim in Quine (1960*a*: chapter 2) that our terms for rocks (or rabbits) could be reinterpreted as terms not for whole, extended, enduring ones, but rather for spatial parts and/or temporal stages thereof, provided compensatory reinterpretations are made of various predicates (beginning with the reinterpretation of 'is the same thing as' as something like 'is spatio-temporally connected to' or simply 'belongs with').

If the nominalist argument was supposed to be that the availability of non-standard and multiple interpretations suffices all by itself to make the case for the impossibility of referring to abstracta, then in view of the foregoing examples pertaining to concreta the argument would be very weak (or rather, would be much too strong, making a case not for nominalism in particular, but for scepticism in general). But the nominalist argument—or at least, the kind of nominalist argument we are considering—is not supposed to be that one. Rather, the argument is that our descriptions are insufficient to pin down the reference of our terms in both the concrete and the abstract cases, but that in the concrete though not the abstract case we have something else that helps, namely our causal relations with the objects we are referring to. Thus Hodes writes of our having an understanding of the 'microstructure' of reference in the concrete case that we lack in the abstract case. 'Adam' refers to Adam and 'Eve' to Eve, and not vice versa, because the relevant causal chains connect our usage of 'Adam' with Adam, not Eve, and our usage of 'Eve' with Eve, not Adam.

The stereotypical anti-nominalist will be quick to point out, however, that considerations of 'causality' don't seem very helpful in attempting to settle whether a given outcropping along the shore counts as part of the Rock of Gibraltar or not. Nor does consideration of the 'microstructure' involved when we look at, point to, and name that famous rock seem to help much in explaining how it is that we manage to refer to the whole thing, and not just the part or stage of it that we see. In the absence of a detailed proposal for a causal theory of reference, it is anything but clear that considerations of causal microstructure will be either sufficient or necessary for the solution or dissolution of the various well-known problems and puzzles roughly and briefly described above. And unless it is, then problems and puzzles about non-standard and multiple interpretation will not show that reference to concreta is secure while reference to abstracta problematic. Or so the stereotypical anti-nominalist will maintain. (See in this connection Wright 1983: 127 ff.)

There is no such thing, however, as the anti-nominalist position on reference. Many proposals have been made, of which a few may be cited here. One type of proposal, aiming to eliminate most indeterminacy, has been developed by Lewis, responding to Putnam, and by Maddy, responding to Kripkenstein. (Compare Lewis (1984), Maddy (1984*b*).) The claim here is that 'the Rock of Gibraltar' refers to the Rock of Gibraltar because that famous rock is a much more 'natural' object than, say, something whose earlier temporal stages coincide with those of the rock but whose later temporal stages drift away from it, or than any other possible candidate. Likewise 'green' means green because green things form a much more 'natural' kind than, say, things in this millennium that are green, and those in the next millennium that are bluish green, or any other possible candidate. Likewise 'sum' means sum because the sum relation is a much more 'natural' one than, say, the relation of sum plus a bonus for large numbers, or any other possible candidate. The general idea is that what determines meaning and reference is not just our usage, but that plus distinctions of 'naturalness' among the things our words might mean or refer to.

Another type of proposal (which may appeal to those who find the notion of 'naturalness' excessively metaphysical) holds in the concrete case that 'the Rock of Gibraltar', for instance, has no one, sole, single, unique referent; that if it is to be credited with any referent at all, then it must be credited with zillions of them, and taken to 'divide its reference' among countless sharply bounded heaps or hunks of matter. But it is proposed that all this does not prevent us from using the term in many true sentences. It is proposed that a sentence involving the term should be counted as true if and only if it is true on all candidate interpretations. Presumably, then, 'the Rock of Gibraltar is primarily Jurassic limestone' is true; whereas some sentence whose truth or falsehood would turn on the exact boundaries of the rock on a scale of centimetres is neither true nor false. The analogous proposal for the abstract case would take terms like 'zero' to vary—or rather, would take systems of terms like 'zero, one, two, . . .' to co-vary—over all the candidate set-theoretic interpretations, but would count as true whatever comes out true on all the candidate interpretations. Thus the whole of standard number theory would come out true, though sentences like 'two is an element of four', whose truth or falsehood would depend on the choice of set-theoretic interpretation, would come out neither true nor false.

A proposal of this type was adumbrated by Benacerraf himself in the speculative closing section of his paper, and is the core of one of the several positions in the philosophy of mathematics that have come to be called **structuralism** (and will provide the starting-point for the construction in Chapter II.C). (A rival theory would hold in the concrete case that vaguely bounded entities like the Rock of Gibraltar are *sui generis*, and not to be identified or confused with any sharply bounded entity. The analogous theory for the abstract case would hold that numbers are objects *sui generis*, not to be identified or confused with any set-theoretic entities. Confusingly, one influential version of this *sui generis* theory in philosophy of mathematics is also called **structuralism**.)

Yet another type of proposal is of special interest, having been adopted by Quine, partly in response to his own examples. It involves an elaboration of the disquotational theory of truth mentioned in the preceding article, according to which the function of the truth-predicate is not for reporting substantive relations of correspondence between language and extra-linguistic reality but rather for the intra-linguistic purpose of undoing quotation. Namely, the theory is extended to the whole family of **alethic** notions, each of which is taken to be governed by a disquotation rule, as illustrated by the traditional paradigms in the adjoining table.

Alethic notions:		
truth	of a sentence	
satisfaction	of a predicate or verb-phrase	by an entity or entities
reference	of a term or noun-phrase	to an entity

Disquotation rules:		
'_____'	is true	if and only if	_____
'_____'	is satisfied	by and only by	whatever _____
'_____'	refers	to and only to	_____
	(if it exists, and otherwise to nothing)		

Traditional paradigms:		
'snow is white'	is true	if and only if	snow is white
'is white'	is satisfied	by and only by	whatever is white
'snow'	refers	to and only to	snow
	(if it exists, and otherwise to nothing)		

The rule for reference makes it completely determinate: 'this rabbit' refers to this rabbit, and not one of its parts or stages; 'zero' refers to zero, and not to one; and so on. (Special provisions have to be made for sentences involving demonstratives, personal pronouns, and so forth.)

The disquotational notions are, however, **local**, applying to our own language, and not **global**, applying to arbitrary languages. The disquotational theory of reference can be extended to foreign languages by the rule:

> a term correctly translated as '_____' refers
>   to and only to _____
> (if it exists, and otherwise to nothing)

But this only makes the notion of reference for foreign languages as determinate as is the notion of correct translation, and Quine maintains (and his examples alluded to above were intended to show) that translation is indeterminate.

Still, Quine holds this indeterminacy of translation generally affects only what he calls radical translation, or translation in the absence of any established cultural contact between two groups of speakers, or any historical connection between their languages. Where there has long been cultural contact and there is a large pool of bilinguals, their practice in passing back and forth between the two languages will be part of the usage constraining translation, and will greatly reduce if not wholly eliminate indeterminacy. Thus it is that we can say:

(i)　　　*'houille'* refers to coal

Again, each generation is in effect bilingual between its own form of the language and that of its parents' generation, and the connection through a chain of generations also reduces or eliminates indeterminacy of translation between past forms of our language and our present form of it. Thus it is that we can say:

(ii)　　　'coal' once referred to charcoal

Philosophers like to consider hypothetical in contrast to actual languages, and often the way in which these are specified (for instance, by contrasting a possible course of historical evolution of our language to the actual one) suffices to reduce or eliminate indeterminacy of translation between a hypothetical form of our language and our actual form of it. Thus it is that we can say:

(iii)　　　'coal' could have referred to snow

Now examples like (i)–(iii) invite the question how 'coal' came to refer, in our actual present language, to what it does, and not to what it might have or once did, or what similar-sounding terms like '*kohl*' in foreign languages do. That question can be raised phylogenetically, as a historical question about the formation of our language in the speech community, or ontogenetically, as a psychological question about the acquisition of our language by young children. And the question can be raised both about terms for the concrete, such as 'coal', and terms for the abstract, such as 'zero'. And it was precisely such questions, in the abstract case, that led the stereotypical nominalist, as depicted by us earlier, to doubts about the possibility of referring to abstracta.

It is therefore of especial interest to consider what a disquotationalist like Quine might make of such questions. Quine has, in fact, devoted a whole book, Quine (1974), to speculation about how the reference of our terms got to be as it is. On examination, the book proves to be concerned solely or mainly with how the usage of our terms got to be as it is: how we came to say 'coal' more often in the presence of a combustible mineral than of winter precipitation, the products of the incomplete combustion of wood, cabbages, or eye shadow, or something else; how we came to count, and to count with the words 'one, two, three, . . .' in that order, and not with some other words, or those words in some other order; and so on. What disquotationalism leaves no room for is any question about how, given our usage, our reference is determined. More generally, disquotationalism rather obviously leaves no room for a strategy that would try to bring into question whether 'coal' really refers to coal, or 'one, two, three, . . .' really refer to one, two, three, . . . respectively, without directly trying to bring into question whether coal, or one, two, three, . . . exist. Any referential argument for nominalism or scepticism must assume a notion of reference more 'heavy duty' than the disquotational.

The literature on causal theories of reference is thin (or more precisely, the literature on causal theories of reference usable for nominalistic purposes will be found to be thin, as soon as one notices that most of the special literature on 'causal theories of reference' provides nothing of use to nominalists). But the literature on non- and un- and anti-causal theories of references (usable for anti-nominalistic purposes) is also thin. The authors proposing the 'naturalness' theory of reference do not claim to have provided more than the beginnings of a sketch of a theory. Comparatively little has been written on 'structuralism' (in the relevant sense, the first as opposed to the second of the two senses noted above). Even on the disquotational theory of reference there are no book-length studies to be cited (as there are for the disquotational theory of truth, for instance, Horwich (1990) and Grover (1992)). Reference is a problem for nominalists if one assumes the burden is on nominalists to enunciate and establish a detailed causal theory of reference; and it is a problem for anti-nominalists if one assumes the onus is on anti-nominalists to enunciate and establish a detailed account of reference to abstracta.

Having now thrice noted the tendency of nominalist arguments and anti-nominalist counter-arguments to reach stalemate over issues of burden of proof, in line with our general aims in this book (as explained at the end of article o.b) we will make no further effort to break through the stalemate ourselves, but rather will leave further reflection on it to the interested reader, as we turn from the negative, destructive side of nominalism to consider its positive, reconstructive side.

3. WHY RECONSTRUAL?

The relationship between the two sides of nominalism is rather curious. Nominalists with reconstructive projects hardly ever fail to cite the destructive arguments, as if the arguments provided motivation for the projects. Yet the arguments do not seem to be of the right form to provide such motivation. They would if they led to the complex conclusion that retaining current theories without seeking alternatives would be unjustifiable, though retaining current theories if alternatives cannot be found would be justifiable. But they do not seem to be of the right form to reach such a conclusion. Rather, they seem to be of such a form that, if they are cogent at all, then they already suffice to establish that retaining current theories would be unjustifiable, regardless of the success or failure of any search for alternatives.

Consider, for instance, the argument that ordinary mathematical and scientific views are rendered dubious by the fact that everything we say about numbers could be reinterpreted as being about certain other abstracta. Surely, if this really does render ordinary views dubious, it must do so whether or not everything we say about numbers could also be reinterpreted instead as being about certain concreta. Or consider the argument that claims that ordinary mathematical and scientific judgements are problematic because they could only be true by a lucky accident. Surely, if ordinary views genuinely are problematic for this reason, they must be so whether or not there are available any alternative views one might adopt to replace them.

It would seem that if one reposes any serious confidence in the arguments in the negative, destructive side of the nominalist literature, one
ought to draw the conclusion that standard science and mathematics are
no reliable guides to what there is. This need not imply any criticism of
science. Or rather, it does imply a criticism of science in the role of a
guide to what to believe, but it is compatible with an appreciation of
science in various other roles, and in particular in that of a guide to what
to do. The claim that science is a fiction is compatible with an appreciation that it is a useful one: the philosopher who rejects science in the
ordinary sense of being unwilling to believe it, may nonetheless 'accept'
science in the special sense of being willing to apply it. Such a philosopher may praise science to the skies as a wonderful and practically
indispensable picture of nature, illuminating and extremely useful—but
just not true.

A contemporary philosopher who was led by doubts about the existence of abstracta such as numbers to adopt such a double attitude towards
science would find ample precedent for it in the views of thinkers
of earlier periods who doubted the existence of unobservable concreta
such as atoms. One such thinker was Ernst Mach (already mentioned in
article 1.b), equally famous for his work in applied science—the Mach
number is named after him—and for his staunch opposition to atomism.
Another was Pierre Duhem, a prominent early contributor to chemical
thermodynamics, and to the history of medieval science, as well as a
resolute sceptic about theoretical physical entities. Yet another was Hans
Vaihinger, today remembered alike as an eminent Kant scholar and founder
of the journal *Kant Studien*, and as the author of the philosophy of '*als ob*'
('as if'), according to which living by fictions is indispensable to us as
creatures who have no hope of knowing reality as it is in itself.

The best-known present-day representative of this tradition is Bas
van Fraassen, whose book van Fraassen (1980) revived scepticism about
atoms and the like under the somewhat odd name of 'constructive empiricism'. The central notions of his philosophy of science are those of **empirical equivalence** and **empirical adequacy**. Roughly and briefly, theories
are empirically equivalent if the expectations about the observable
one would form if one immersed oneself in the one theory are the same
as those one would form if one immersed oneself in the other; and a
theory is empirically adequate if the expectations about the observable
one would form if one immersed oneself in the theory would be correct.
Where the orthodox assert a theory T, van Fraassen asserts that the
theory T is empirically adequate. One may call this assertion T_0 a 'theory',

and if so T_0 is a 'theory' empirically equivalent to T. But van Fraassen sees no need to develop any empirically equivalent alternative to T less immediately parasitic on T than is this T_0. He makes no attempt to develop any detailed theory capable of yielding the various predictions about the concrete implied by T, while avoiding the involvement with unobservables present in T.

(For that matter, he makes no attempt to develop any detailed language capable of expressing the various predictions about the concrete implied by T, while avoiding the involvement with unobservables present in the language of T. When a chemist predicts that one will soon smell the characteristic stench of H_2S, which is to say, of the substance composed of molecules each consisting of two atoms of hydrogen and one of sulphur, van Fraassen might say that one will soon smell the characteristic stench of the substance which *according to chemists* is composed of molecules each consisting of two atoms of hydrogen and one of sulphur. Or he might just say that one will soon smell the characteristic stench of H_2S and leave the italicized phrase tacit: he might just say that one will soon smell the characteristic stench of H_2S while 'bracketing' any atomist implications of this mode of expression, just as we all say that one will soon see the sun set while 'bracketing' any geocentrist implications.)

In short, the variety of sceptical empiricism just roughly and briefly described undertakes no positive, reconstructive programme. (A less rough and brief account of van Fraassen's empirical scepticism can be found in Rosen (1994).) Now the negative, destructive arguments of nominalism are not supposed to lead to doubts about unobservables in general—though we noted in section 2 that at several points they do seem to threaten to do so—but only to doubts about abstracta in particular. For this reason, the typical nominalist will not wish to adopt an attitude towards science *identical with* van Fraassen's. But one can easily imagine an attitude *analogous to* van Fraassen's, but which restricts its doubts to numbers and the like, not atoms and their ilk. The central concepts would be those of **nominalistic equivalence** and **nominalistic adequacy**, defined like empirical equivalence and empirical adequacy, but with 'about the concrete' replacing 'about the observable'. Presented with a standard theory, van Fraassen's nominalistic counterpart would make no attempt to develop a nominalistic reconstrual or reconstruction of it but would be content with the 'theory' consisting of the assertion that the standard theory is nominalistically adequate, that the world is *in all concrete respects as if* the theory were true. This is the type of position that we earlier (in article o.b) called 'instrumentalist nominalism' and that one of us elsewhere

(Rosen 1992: chapter 3, where a less rough and brief account can be found) has by analogy with van Fraassen's terminology called 'constructive nominalism'.

Against van Fraassen's view it is sometimes objected that, rejecting standard hypotheses about unobservable causes of observed phenomena, he fails to provide any causal explanation of empirical data; but the nominalistic analogue of van Fraassen's view would clearly not be open to any such objection. Against van Fraassen's view it is also sometimes objected that, without the assumption that standard theories are true, one is left with no good explanation of why they are empirically adequate; and the nominalistic analogue of van Fraassen's view would clearly be open to an analogous objection. However, the objection to van Fraassen invites an immediate retort from the opponents of unobservables: if there is anything to the negative, destructive arguments against unobservables, then even the assumption that the standard theories are true does not provide a good explanation. For if there is anything to those arguments, then the standard theories are just bad, and one can't give a good explanation by assuming a bad theory. Clearly an analogous retort could be made by the opponents of abstracta.

The reconstructive nominalist seems to face a dilemma. On the one hand, if the negative, destructive arguments are taken seriously, there would seem to be a need for some further argument in order to show that a positive, reconstructive project is necessary. Namely, there would be a need for some cogent objections to mere 'instrumentalist nominalism' or 'constructive nominalism'. On the other hand, if the negative, destructive arguments are not taken seriously, if they are cited only to pique interest in nominalism, there would seem to be a need for some argument to show that the success of a positive, reconstructive project would be sufficient to establish nominalism. For after all, to say that such a project has succeeded is only to say that there is a nominalistic alternative to standard scientific theory that could be adopted in its place. But should it be? Some further argument would seem to be needed to bridge the gap between 'could' and 'should' here. Thus whether one thinks the negative, destructive arguments to be powerful or feeble, there would seem to be something more needed to motivate the positive, reconstructive projects. Yet it is very difficult to locate in the writings of the reconstructive nominalists any sustained argumentation either for the necessity of their projects (assuming the negative arguments are conclusive and establish nominalism) or for the sufficiency of their projects to establish nominalism (assuming the negative arguments are no more than suggestive).

This, we suspect, is because reconstructive nominalists generally present themselves as replying to a certain specific and highly influential anti-nominalist argument, and because that anti-nominalist argument makes a major concession to nominalism, essentially the concession that if nominalistic alternatives to standard scientific theories could be developed, then they should be adopted. The anti-nominalist argument in question can be found in scattered passages in the post-nominalist writings of the ex-nominalist Quine, but the *locus classicus* is the booklet Putnam (1971). Eliding certain differences between Quine's version and Putnam's, it is usually called 'the Quine–Putnam **indispensability** argument'. Roughly and briefly put, it amounts simply to the claim that we should believe in abstract entities, but only because nominalistic alternatives to standard scientific theories cannot be developed.

We have just said that making the case for abstracta depend on the impossibility of dispensing with them in science means making a major concession to nominalism, namely, the concession that if it were possible to dispense with abstracta it would be desirable to do so. But it also represents a concession on the part of the reconstructive nominalists that they feel obliged to respond to such indispensability claims. That they have felt such an obligation shows something important about the present state of (Anglophone) philosophy. In the climate of philosophical opinion prevailing in some earlier eras (or in that prevailing today in some other parts of the world), the mere fact that a philosophical thesis appears irreconcilable with science would hardly have been considered relevant. That nominalists feel an obligation to respond to indispensability claims testifies to a genuine rarity: a near-consensus among contemporary (Anglophone) philosophers, namely, the near-consensus that current scientific theory has a prima-facie if not indefeasible claim on our belief.

One may distinguish two ways in which such philosophical deference to science might be supported. Some philosophers have regarded the science of their day as authoritative because they took themselves to have provided an external, philosophical foundation for it. Other philosophers regard science as credible, and as imposing a constraint on what can be taken to be credible in philosophy, not because they take themselves to have provided any external support for its theories, but simply because they recognize no better kind of support for a theory than the kind of support for theories internal to science itself. This latter stance is characteristic of what we earlier called 'naturalized' as opposed to 'alienated' epistemology, and something very like it was presupposed by Quine and Putnam, the main proponents of indispensability arguments.

Thus in Putnam (1971: §8) we read:

The fictionalist concedes that the predictive power and 'simplicity' . . . are the hallmarks of a good theory, and that they make it rational to accept a theory, at least 'for scientific purposes'. But then . . . what further reason could one want before one regarded it as rational to believe a theory?

Putnam here makes no claim to have provided an external, philosophical foundation for science refuting the external, philosophical critiques of science put forward by Vaihinger or Duhem or Mach; nor does he even claim to have shown that the views of such thinkers are internally incoherent. He simply takes his stand with science. Inasmuch and in so far as, in putting forward indispensability considerations in opposition to nominalism, he presupposes rather than argues for 'naturalization' or 'naturalism', it is somewhat misleading to speak of an 'indispensability argument'. But 'argument' or not, the indispensability considerations have been something to which nominalists have felt obliged to respond, because they, too, for the most part profess to be adherents of 'naturalization' or 'naturalism'.

It is to be stressed that naturalism in this sense is not to be confused with crass 'scientism', according to which no answer to any question is credible unless sanctioned by those with credentials as professional scientists. For in the first place, the various forms of enquiry we call science do not speak to every question. There is no such thing as political science, for instance; yet no professed naturalist maintains that we should abstain from having political views. And in the second place, science is not a closed guild with rigid criteria of membership. Philosophers professing naturalism often do contribute to debates in semantical theory or cognitive studies or other topics in the domain of linguistics or psychology, even though they are not officially affiliated with a university department in either of those fields. In principle nothing would bar such philosophers from participating in discussions on topics in the domain of chemistry or geology, though in practice they seldom do. The naturalists' commitment is at most to the comparatively modest proposition that when science speaks with a firm and unified voice, the philosopher is either obliged to accept its conclusions or to offer what are recognizably scientific reasons for resisting them.

We noted earlier that the anti-nominalist philosophers who first put forward the indispensability argument effectively conceded to their opponents that if abstracta could be eliminated from science, then they should be. It may now be added that in making this concession to nominalism they did not take themselves to be deviating from their professed naturalism and making a concession to alienated epistemology. Rather, they seem to have assumed that if a nominalistic alternative to current scientific theory were produced, it would automatically be superior scientifically. If this is so, then nominalists with reconstructive projects can dispense entirely with the equivocal support provided by the questionable destructive arguments we surveyed in section 2, and rest the case for their projects simply on their scientific merits.

But is it indeed so that if a nominalistic alternative to current scientific theory were produced, it would automatically be superior scientifically? Or more to the point, is it indeed so that the nominalistic alternatives to current scientific theories that have been produced actually are superior scientifically? That is a question we will defer to the end of this book (in section III.C.1), since it is a question best postponed until we have completed nominalistic reconstructive projects before us. In the meantime we note the following: if the question has to be answered in the negative— if the scientific merits of the nominalistic reconstruals or reconstructions are not such as to permit the case for their superiority over current scientific theory to be made on purely scientific grounds—then the nominalist will have to fall back on destructive arguments of the kind we surveyed in section 2, unless others can be produced. We have suggested that the destructive arguments surveyed are not unquestionable, and that their form is such that the support they give to reconstructive nominalism is not unequivocal. In line with our general aims in this book (as explained at the end of article 0.b), we will make no effort to imagine what other arguments might be produced, but rather will leave speculation on that topic to the interested reader.

B

A Common Framework for Strategies

o. OVERVIEW

Let us now accept as 'a hypothetical statement of conditions for the construction in hand' that the kind of technical science in which advanced mathematics is applied is to be reconstrued nominalistically. In the literature, indicating a strategy for doing so is usually accomplished by indicating how a formalized version, written in an artificial language based on a symbolic logic, of a standard theory is to be paraphrased into a formalized version of a nominalistic theory. That is why some knowledge of how to read formulas of such artificial languages must (as indicated in article I.A.o.a) be presupposed in this book. While in the literature there is general agreement in presenting strategies in a formal (or semiformal) framework, differing frameworks are adopted by different authors for their differing strategies. For purposes of comparative study, it will be desirable to set up a common framework, and to suggest one is the aim of this chapter.

Actually, there will be very few purely symbolic formulas considered, for the process of **formalization**, of paraphrase from a natural language like English into an artificial language based on a symbolic logic, has several stages, and it is very seldom necessary to go all the way to the last stage. The first stage is what Quine calls **regimentation**. At this stage, using skills taught in introductory-level logic courses, one paraphrases ordinary English with its vast range of grammatical and logical constructions and operations into a stylized English that Richard Jeffrey calls 'Loglish', with a very limited range of such constructions and operations. Thus:

(i)   A part of a part of a thing is a part of that thing

might become:

(ii)   For any x, for any y, for any z, if y is a part of x, and z is a part of y, then z is a part of x

Next comes **symbolization**, the mechanical transcription of words into symbols. It has two aspects. One is the transcription of non-logical vocabulary. For instance 'is a part of' may be replaced by some single symbol. Here it will be convenient to use the symbol '$\propto$'. With this notation, (ii) would become:

(iii)   For any x, for any y, for any z, if $y \propto x$, and $z \propto y$, then $z \propto x$

The other is the transcription of the logical vocabulary. With the notation used here, (ii) would become:

(iii') $\forall x \forall y \forall z((y$ is a part of $x \land z$ is a part of $y) \to z$ is a part of $x)$

Transcribing *all* vocabulary produces purely symbolic formulas. For instance, (ii) would become:

(iv) $\forall x \forall y \forall z((y \propto x \land z \propto y) \to z \propto x)$

But we will seldom go this far.

Standard Logical Apparatus

Symbol	Name	Reading
~	Negation	[it is] not [the case that]
$\land$	Conjunction	[both] . . . and . . .
$\lor$	Disjunction	[either] . . . or . . .
$\to$	Conditional	if . . . [then] . . .
$\forall$	Universal Quantification	for all . . .
$\exists$	Existential Quantification	for some . . .

Since the symbolism for logical operators used varies from introductory textbook to introductory textbook, we list ours in the adjoining table. Initially, all the languages and theories considered here will be ones based on **standard** logic which has only the above operators. (It is also called **classical** logic when contrasted with various **restricted** logics, or **elementary** logic when contrasted with various **extended** logics.)

Since the terminology used for notions pertaining to logical formulas varies from introductory textbook to introductory textbook, we will do well to indicate ours here: in a language based on standard logic there are, in addition to the logical predicate = of identity common to all such languages, certain **primitive** non-logical predicates peculiar to the language in question, each with a fixed number k of places. A k-place predicate followed by k variables constitutes an **atomic** formula. Other, **molecular** formulas are built up from atomic ones using the classical operators. All the variables in an atomic formula $Px_1 \ldots x_k$ are **free**; the free (respectively, bound) variables in the negation of a formula or conjunction or disjunction of two formulas are just the free (respectively, bound) variables of the formula(s) negated or conjoined or disjoined; the free variables of a quantification $\forall x Q$ or $\exists x Q$ are those of the formula quantified except for the variable x, which is **bound**. A formula is **open** or **closed** according as it has some free variables or has only bound variables. Initially, all the languages considered here will be ones having only finitely many non-logical primitives. All the theories considered here will be ones having, in addition to the logical axioms for the various operators and for identity, only finitely many non-logical axioms, or at worst, only finitely many non-logical axioms and finitely many non-logical axiom schemes. Here a **scheme** is a rule to the effect that all formulas of a certain form are to count as axioms. For instance, the logical axioms for identity consist of one single axiom, $\forall x(x = x)$, and one scheme of axioms, according to which for every formula F the following is an axiom:

$$\forall x \forall y (x = y \rightarrow (F(x) \rightarrow F(y)))$$

It will be convenient to work with a **two-sorted** language, which is slightly different from the ordinary or one-sorted languages found in most introductory textbooks and in most of the literature. Such a language has two sorts of variables, one $x, y, z, \ldots$, called **primary**, ranging over one sort of entity, also called primary; and another $X, Y, Z, \ldots$, called **secondary**, ranging over another sort of entity, also called secondary; and each primitive is assigned not just a fixed total number k of places for variables, but rather a fixed number m of places for variables of the first sort, and a fixed number n of places for variables of the second sort. Thus primitives may be classified as **primary** ($n = 0$), **mixed** ($m, n > 0$), or **secondary** ($m = 0$); and formulas of the language may be classified as **primary** if they contain only primary primitives, **secondary** if they contain only secondary primitives, and otherwise as **mixed**.

Two-sorted languages are convenient for the following reason. Scientific theorizing, especially in sophisticated physics, which nominalists generally recognize as posing the greatest challenge, typically involves associating with some original structure of concrete, physical entities a corresponding image structure of abstract, mathematical entities. Theorems about the latter structure, taken together with the assumption of correspondence, imply predictions about the former structure. In the usage of mathematicians, proving theorems specifically in order to be able to use them in this way is called doing **applied** mathematics, and proving theorems without this specific intent is called doing **pure** mathematics. In the usage of philosophers, however, the theorems proved in either case are **purely** mathematical, since they mention only mathematical entities, whereas the assumption of a correspondence between a physical and a mathematical structure is a **mixed** mathematico-physical assumption, since it mentions both mathematical and physical entities. The theory for which a nominalistic reconstruction is sought is the one consisting of both the assumption of correspondence and whatever purely mathematical axioms are needed to deduce whatever purely mathematical theorems are applicable or useful for deducing predictions in the way indicated. It will be convenient, therefore, to take a scientific theory to be formalized in a two-sorted language in which the primary variables are supposed to range over the concrete, physical entities, in some concrete, physical structure and the secondary variables over abstract, mathematical entities in some abstract, mathematical structure. The secondary axioms of the theory will be the purely mathematical axioms just mentioned, while the mixed axioms will assert how the mathematical structure is assumed to correspond or supposed to relate to the physical structure.

At the outset in a positive nominalist project, it would be an advantage to be able to assume as few mathematical or secondary entities and axioms as feasible. For the fewer there are at the outset, the easier it will be to eliminate them in the end. Under the rubric of mathematical entities fall both all pure mathematical entities, such as numbers or sets of numbers, mentioned in the purely mathematical axioms, and all impure mathematical entities (if any), such as sets of physical entities, mentioned in the assumptions about the correspondence between the physical and the mathematical. Even just considering only the pure mathematical entities for the moment, sophisticated physical theories make use of a great variety of exotic mathematical entities, from the 'Riemannian manifolds' of differential geometry used in general relativity to the 'Hilbert space' of functional analysis used in quantum mechanics. Fortunately, all these greatly varied mathematical entities can be represented by sets, and the greatly varied branches of mathematics from which they come be reduced to set theory, by a process sketched in article 1.a. Unfortunately, the usual axioms of set theory, the ones semi-officially taken by mathematicians today as the framework for mathematics, assume a vast hierarchy of higher and higher ranks of sets. Fortunately again, there is a consensus of expert opinion that the mathematics needed for applications can be developed in a much more modest framework, that of mathematical analysis, whose content is explained in article 1.b. Moreover, it turns out that a plausible, though admittedly not irresistible, case can be made out for the assumption that no mixed mathematical entities will be needed after all, for reasons explained in article 1.c.

Altogether section 1 gives an account of what has to be gotten rid of in a nominalistic reconstruction. Getting rid of (assumptions about) entities of one sort by means of (assumptions about) entities of another sort is called **ontological reduction** (or **elimination**). In the case of the reduction of other mathematicalia to sets in article 1.a, or to real numbers in article 1.b, the notion of reduction or elimination in question is the simplest conceivable. Each entity X of the sort to be eliminated is assigned a surrogate or proxy entity x of the sort to which the eliminated entities are to be reduced, this x being said to **represent** or **code** X. Predicates applying to the entities to be eliminated are reinterpreted as applying to their representatives or codes, and quantifications over the entities to be eliminated are reinterpreted as quantifications over their representatives or codes. For purposes of nominalistic reconstrual, a more sophisticated notion of reduction (or elimination) will be needed, and is expounded in section 2 (informally in the optional quasi-historical digression article 2.a, semi-formally in article 2.b). The proof of the main claim about this notion of reduction is given in outline in articles 3.a–3.c (with a few more details related to the optional semi-technical appendix article 3.e). (This notion of reduction is contrasted with others, generally agreed to be of no philosophical significance, in section 4, another optional semi-technical appendix, which presumes some familiarity with at least semi-popular accounts of some results from intermediate-level logic.)

Finally, article 3.d recapitulates for ready reference as much of the work of this chapter as needs to be remembered in later ones. From the common starting-point there indicated, the differing strategies outlined in the succeeding chapters move rapidly off in widely differing directions.

1. INPUT TO BE ELIMINATED

a. Pure Sets

In a typical introductory textbook on set theory—we have a couple before us as we write—roughly the first half will be devoted to set theory as a framework for the rest of mathematics, and the second half to set theory as a special branch of mathematics in its own right. And typically the reconstruction of mathematics within a set-theoretic framework will be presented in the first half of the book in roughly the following sequence of steps.

To commence, there will be an exposition of the axioms of the usual system of set theory, called **ZFC**, which is the end product of the set-theoretic or **synolist** tradition of Georg Cantor, Ernst Zermelo, and Abraham Fraenkel. (This typically is the topic chapter 1 of a textbook.) The details of their formulation will not be important here. Suffice it to say that they are partial descriptions of a hierarchy of levels or ranks of sets of the following kind. At rank zero come individuals. At rank one come sets all of whose elements are individuals, beginning with the empty set { }. At rank two come sets all of whose elements are individuals or sets of the first rank, beginning with the set {{ }}. At rank three come sets all of whose elements are individuals or sets of the first rank or sets of the second rank, beginning with the set {{{ }}} and the set {{ }, {{ }}}. After all finite ranks comes a first infinite rank, rank omega, where come sets all of whose elements are individuals or sets of finite rank, beginning with the sets:

$$\{\{ \}, \{\{ \}\}, \{\{\{ \}\}\}, \ldots\}$$
$$\{\{ \}, \{\{ \}\}, \{\{ \}, \{\{ \}\}\}, \ldots\}$$

Since sets come in higher and higher ranks, with the elements of sets of higher ranks coming from lower ranks, there is no set of all sets, and no set is an element of itself. The famous **Russell paradox** about the set of all sets that are not elements of themselves does not arise, because this would be the set of all sets, and there is no such set.

To continue, there will come some generalities about how entities ostensibly of other sorts are to be represented by sets. (This typically is the topic of chapter 2 of a textbook.) Notably, an ordered pair (a, b) is represented by the set $\{\{a\}, \{a, b\}\}$, a two-place relation R by the set of ordered pairs $\{(a, b) \mid a \text{ is } R\text{-related to } b\}$, and a function f is represented by the relation of argument to value. All this will be used in the reconstruction of the traditional number systems within set theory, which is what comes next. This is undertaken in stages. (It typically takes up several chapters in a textbook.)

The first stage is the reconstruction of the system consisting of the natural numbers with the natural order on them, including the proof of its basic laws, the **progression** axioms:

> there is a $\leq$-least number, zero 0
> for any number ξ there is a $\leq$-least number $\leq$-greater than ξ,
>   the successor ξ^+
> for any set X, if $0 \in X$ and if $\xi^+ \in X$ whenever $\xi \in X$,
>   then $\xi \in X$ for all numbers ξ

This last is the **induction** axiom. The natural numbers may be constructed from sets or assigned set-theoretic surrogates or proxies either in Zermelo's way:

$$0 = \{ \} \qquad 1 = \{0\} = \{\{ \}\} \qquad 2 = \{1\} = \{\{\{ \}\}\}$$
$$3 = \{2\} = \ldots \qquad \ldots$$

or in John von Neumann's:

$$0 = \{ \} \qquad 1 = \{0\} = \{\{ \}\} \qquad 2 = \{0, 1\} = \{\{ \}, \{\{ \}\}\}$$
$$3 = \{0, 1, 2\} = \ldots \qquad \ldots$$

since the same laws are provable with either type of definition.

The second stage is the proof of the existence of a unique function, addition, satisfying the recursion equations, $\xi + 0 = \xi$ and $\xi + \upsilon^+ = (\xi + \upsilon)^+$, and the deduction from these equations of the usual laws, beginning with associativity, $\xi + (\upsilon + \zeta) = (\xi + \upsilon) + \zeta$, and similarly for multiplication.

The third stage is the reconstruction of the rational number system, defining addition and so forth and deducing associativity and so forth for rational numbers from the corresponding notions and laws for the natural numbers. The construction exploits the fact that, intuitively speaking, every rational number can be canonically represented as $\pm\xi/\upsilon$ for some natural numbers ξ, υ having no common factors.

The fourth stage is the reconstruction of the real number system. There are several constructions of the reals from the rationals, the best known being that due to Richard Dedekind and that due to Georg Cantor. Intuitively speaking, one construction represents a real number X by the set of all rational numbers less than X; another represents X by the set of all rational numbers that **approximate** X, where a rational $\pm\xi/\upsilon$ in canonical form is said to approximate a real X if it differs from X by less than $1/\upsilon$. The same laws are provable with either type of definition, including in particular the crucial **continuity** law distinguishing the reals from the rationals:

> for any set Ξ of real numbers,
>   if there is at least one real number $X \in \Xi$
> and at least one real number $Y \notin \Xi$
>   and if $X < Y$ for every $X \in \Xi$ and $Y \notin \Xi$
> then there is a real number Z such that
> $X \in \Xi$ for all $X < Z$ and $Y \notin \Xi$ for all $Y > Z$.

The account of the reduction of other mathematical objects to sets typically concludes at this stage, as it can since in rigorous university-level textbooks and research monographs and papers the other objects of mathematics, up through and including the Riemannian manifolds and Hilbert space alluded to in section o, are generally constructed from sets, relations, and functions from the natural, rational, and real number systems. (For purposes of the foregoing construction, it is not necessary to assume the set of individuals non-empty. The **pure** sets as opposed to **impure** sets are those in the hierarchy over the empty set of individuals, and they include all those used in the reconstruction of the traditional numbers systems.)

The transition from the study of set theory as a framework for the rest of mathematics to the study of set theory as a branch of mathematics in its own right begins with some distinctively set-theoretic notions and results that are often used in other branches of mathematics (and often included in introductory- to intermediate-level logic courses). Above all, this includes the basic definition of Cantor that two sets A and B have the same **cardinal number** if there is a **bijective** relation R between the elements of the one and the elements of the other, where bijectivity means that for every $a \in A$ there exists a unique $b \in B$ such that a is R-related to b, and for every $b \in B$ there exists a unique $a \in A$ such that a is R-related to b. Also included are the basic theorems of Cantor about transfinite cardinals, the cardinal numbers of infinite sets. The cardinality of the set of natural numbers is the smallest transfinite number and is called $\aleph_0$, sets of this smallest infinite cardinal being said to be of **countable** size. Examples, according to Cantor's theorems, are the sets of integral or rational numbers, and that of finite sequences from a finite alphabet of symbols. The cardinality of the set of real numbers is a larger transfinite number and is called $\mathfrak{c}$, sets of this cardinal being said to be of **continuum** size. Examples, again according to Cantor's theorems, are the set of complex numbers, or those of sets of natural or rational numbers, or that of infinite sequences from a finite alphabet of symbols. (In a textbook, these results may be developed bit by bit in the course of the construction of the number systems.)

Though set theory has the reputation of being full of proofs that purport to establish the existence of a mathematical entity with a certain mathematical property without specifying any definite such entity, this reputation is not wholly deserved. Certainly the proofs just alluded to do provide specific codings of rational numbers or finite sequences from a finite alphabet by natural numbers, and of complex numbers or sets of natural numbers by real numbers. A related theorem states that the set of countable subsets $\{a_0, a_1, a_2, \ldots\}$ or countable ordered sequences

$(a_0, a_1, a_2, \ldots)$ from a set of size $\mathbf{c}$ has size $\mathbf{c}$. Implicit in the proof is a coding of such a set or sequence of real numbers by a single real z, the one having the digits of the decimal expansion of a_0 in the odd-numbered places of its decimal expansion, those of a_1 in the places divisible by two but not by four, those of a_2 in the places divisible by four but not by eight, and so on. For instance:

$$\{.000000\ldots, \quad .111111\ldots, \quad .222222\ldots, \quad .333333\cdots,$$
$$.444444\cdots, \ldots\}$$

is represented by the real:

$$.0102010301020104010201 03 \ldots$$

Then, iterating, countable sets or sequences of countable sets or sequences, countable sets or countable sets of those, and so on, can be represented by single reals.

Another theorem of the kind indicated states that the set of all open and closed sets of real numbers, and the set of all continuous functions from and to the real numbers have size $\mathbf{c}$. Here an open set Ξ of real numbers is such that for any $X \in \Xi$ there are $A < X$ and $B > X$ such that $Y \in \Xi$ for all Y with $A < Y$ and $B > Y$; a closed set is one such that for any X, if for every $A < X$ and $B > X$ there is a $Y \in \Xi$ with $A < Y$ and $B > Y$, then $X \in \Xi$; and a continuous function f is one such that for any open set Ξ the set of X such that $f(X) \in \Xi$ is also open. The proofs provide codings of open sets and of continuous functions by sets of ordered pairs of rational numbers, and hence in view of the preceding paragraph ultimately by real numbers. Moreover, the coding can be extended to wider classes of sets and functions, through and indeed beyond what are known as the **Borel** sets and functions—enough for the functional analysis needed for quantum mechanics, and more than enough for the differential geometry needed for general relativity.

Set theory does, however, partly deserve the reputation mentioned above, especially on account of the role in it of the axiom of **choice**. The details of its formulation will not be important here, though the equivalence of various formulations is the last of the topics from set theory taught to working mathematicians (and is always included in textbooks).

General pure set theory as a subject in its own right (the topic of the latter half of a typical textbook) begins with **Cantor's Theorem**: according to the **power** axiom, for every set I the power set $\wp(I)$ or set of all subsets of I exists; and the theorem in question generalizes the result that $\aleph_0 < \mathbf{c}$ to the result that for any given infinite cardinal (say that of a given infinite set I) there is a larger one (namely that of the power set $\wp(I)$). (The **continuum hypothesis** is that there is no cardinal between $\aleph_0$ and **c**.) This unending series of cardinals is the object of investigation in general pure set theory.

b. Pure Numbers

Pure general set theory has not yet, however, shown itself relevant to much of the rest of mathematics, let alone to physics. The set theory needed for $99^{44}/_{100}$ per cent of pure mathematics can be developed in much weaker systems than ZFC. One such system involves a **stratified** rather than a **cumulative** hierarchy, meaning a hierarchy in which the elements of a set must come from the immediately lower level, rather than from arbitrary lower levels. (This hierarchy has only finite, not infinite levels of sets, and the level of individuals at the bottom must be assumed infinite and not empty.) Under a different terminology (**class** vs. **set** for the collections, **member** vs. **element** for their constituents, **type** vs. **rank** for the levels of the hierarchy, **simplified theory of types** for the theory) this theory is the end product of the **logicist** tradition of Gottlob Frege, Bertrand Russell, and Frank Plumpton Ramsey.

The mathematics needed for virtually 100 per cent of known applications can be developed in weaker systems still, such as ZFC⁻, the result of dropping the power axiom from ZFC. The sets whose existence is provided for by this theory, the **hereditarily countable** sets (sets that are themselves countable, whose elements are all countable, the elements of whose elements are all countable, and so on), can all be coded by real numbers. The mathematics needed for known applications can indeed be developed in a theory in which the only mathematical entities are real numbers. We turn next to the consideration of this theory.

It has been the received view and expert opinion among competent logicians since the 1920s that the mathematics needed for applications can be developed in a theory known as **mathematical analysis**, in which the only entities mentioned are real numbers. The language of analysis has variables $X, Y, Z, \ldots$ for real numbers, and primitives $\leq$, Σ, Π, I for the **order, addition, multiplication,** and **integrity** relations:

> X is less than Y
> the sum of X and Y is Z
> the product of X and Y is Z
> X is non-negative and integral

Various other notions can be defined in terms of these (beginning with identity, which therefore need not be taken as a primitive, since real numbers are identical if and only if neither is less than the other). The theory of analysis has a finite list of basic **algebraic** axioms for order, sum, and product, beginning with the associative law of addition (also of multiplication). It also has a finite list of appropriate axioms for integrity, whose details will be suppressed here. As for **continuity**, it cannot be formulated as a single axiom as was done in article 1.a, since that formulation mentions sets of real numbers. Instead, there is an axiom scheme of continuity, with an axiom for each formula R:

> if there is some X such that $R(X)$ and
>   there is some Y such that not $R(Y)$
> and if $X < Y$ for every X such that $R(X)$ and
> Y such that not $R(Y)$
> then there is a Z such that $R(X)$ for all $X < Z$,
>   and not $R(Y)$ for any $Y > Z$

The list of instances of this scheme for formulas of the language of analysis does not exhaust the content of the law of continuity as stated in terms of sets in article 1.a: more can be proved from the standard axioms ZFC of set theory about the real numbers as standardly reconstructed in set theory, than can be proved about real numbers just from the axioms of analysis. However, to repeat, all the mathematics needed for applications can be reconstructed using only the weaker axioms of analysis.

c. Impure Mathematicalia

The process—or rather, the numerous and varied processes—of application of mathematics to science have never been analysed by logicians with the same thoroughness as has been the pure mathematics needed for scientific applications. In the absence of such an analysis, no claim about what impure mathematical entities would be needed can be compelling. Some claims can, however, be made plausible. Consider first the measurement of continuous quantities.

On the most straightforward approach, this would involve predicates like 'X is the mass of x', involving physicalia x and impure numbers or numbers-with-units X. However, by reconstruing or reparsing 'the mass of x is two grams' as 'the mass-in-grams of x is two', impure numbers can be avoided: only pure numbers are required for the measurement of intensive magnitudes (mass, charge) or extensive magnitudes (length, area, volume, duration), or positional magnitudes (spatial and temporal coordinates).

A mixed primitive M involving physicalia and pure real numbers will be called a **measurement** primitive, and a mixed formula $M(x, X)$ will be said to express a **measurement** notion in the language of a given theory if it can be deduced from the axioms of the theory that:

for every x there exists a unique real number X such that $M(x, X)$

The archetypal example would be 'X is the mass of x' (in grams or some other arbitrary but fixed, though here unspecified, units), or 'X measures how massive x is'. Note that if $M_1, \ldots, M_k$ express measurement notions, and analysis is assumed, then $M(x, X)$ given by:

X represents a k-tuple $(X_1, \ldots, X_k)$ such that
$M_1(x, X_1)$ and $\ldots$ and $M_k(x, X_k)$

also expresses a measurement notion: since k-tuples of real numbers can be represented by single real numbers, taking a whole **profile** of measurements can be regarded as taking a single measurement.

It seems plausible that, if one starts with a sufficiently comprehensive portion of overall scientific theory, there will be numerous enough and varied enough measurement notions definable so that for a sufficiently comprehensive profile of them a kind of **exclusion principle** or result will be acceptable:

for every X there is at most one x such that $M(x, X)$

That is, no two physical entities have the same measurement profile; or in other words, no two distinct physical entities have exactly the same spatio-temporal position, mass, charge, and so on. The exclusion principle, if not a conceptual truth in the strictest sense, is at least compatible with or implied by a wide range of physical theories.

Consider now the counting of discrete units by natural numbers. Take first the counting of the number of real numbers X satisfying some condition $R(X)$. One would like to be able to associate with any such formula a **counting** formula $R^{\#}$ expressing:

[Y is a non-negative integer and
   there are only finitely many X such that $R(X)$ and]
the number of X such that $R(X)$ is Y

The most obvious way to do this would be to assert the existence of a bijective relation Ξ between the set of X such that $R(X)$ and the set of Z such that Z is non-negative, integral, and less than Y. This would, however, involve mentioning a pure mathematical entity Ξ that is not a real number. That can be avoided by asserting instead the existence of a real number W coding such a Ξ.

Take next the counting of the physical entities x satisfying some condition $Q(x)$. One would like to be able to associate with any such formula a counting formula $Q^{\#}$ expressing:

> [Y is a non-negative integer and
>   there are only finitely many x such that $Q(x)$ and]
> the number of x such that $Q(x)$ is Y

The most obvious way to do this would be to assert the existence of a bijective relation Ξ between the set of x such that $Q(x)$ and the set of Z such that Z is non-negative, integral, and less than Y. This would, however, involve mentioning an impure mathematical entity Ξ.

But assuming an exclusion result as above for some measurement notion M, one can take the required $Q^{\#}$ simply to be $R^{\#}$ where R is 'there is an x such that $M(x,\ X)$ and $Q(x)$'. That is, one need not, in order to express notions pertaining to counting, assume impure mathematical entities.

And more generally, assuming an exclusion principle as above, if physical objects can be represented by real numbers, then countable sets, countable sets of countable sets, countable sets of countable sets of countable sets, and so forth, of physical objects or of physical objects and real numbers, can be represented by countable sets and so forth of real numbers, and hence by single real numbers. So it seems plausible that no impure mathematical entities, and no mathematical entities other than real numbers, will be needed. In a first essay at a comparative survey of nominalistic reconstructive projects, it seems reasonable to make such simplifying assumptions.

2. METHOD OF ELIMINATION

a. *Contextual Reduction*

Before the formal presentation of the method of reducing (theories about) abstracta to (theories about) concreta used in most of the strategies to be surveyed in this book, some informal discussion may be helpful. (It is only helpful, not indispensable; the present article is an optional quasi-historical digression, and the reader may skip ahead to article 2.b.)

The simplest and most direct method of reducing (a theory about) one sort of entity to (a theory about) another sort of entity, is **objectual** or thing-by-thing reduction: each X of the former sort is assigned an x of the latter sort as a proxy or surrogate, an understudy taking on its role, an impersonator assuming its identity. Words referring to an X can then be reinterpreted as referring to the x representing it. Predicates applying to Xs can then be reinterpreted as predicates applying to the xs representing them. Quantification over Xs (universal assertions about all of them and existential assertions about some of them) would be reinterpreted as quantifications over the xs representing them. This is the method involved in the reductions of one sort of abstracta to another we have considered so far (the reduction of other mathematicalia to sets in article 1.a, and of hereditarily countable sets to real numbers in article 1.b). To make explicit what the method requires, one needs to assume or define a notion x **represents** X relating entities of the one sort to those of the other, and one needs to assume or deduce:

(i)   every x represents at most one X
(ii) every X is represented by at least one x
(iii) every X is represented by at most one x

 The method also has some limited utility in reducing abstracta to concreta, if one is concerned with a narrow enough range of abstracta and prepared to assume a broad enough range of concreta. For instance, biological species, considered as the characters that living organisms that are equivalent in the sense of being biologically conspecific or species-mates thereby have in common, can be reduced to conglomerates of physical bodies, if these are assumed. Namely, each species X can be represented by the conglomerate x of all living organisms of that species, the mass of all their tissue. The mass of all horse-flesh can serve as a surrogate for *Equus caballus*, the mass of all pork-on-the-hoof can go proxy for *Sus scrofa*, and similarly for other species. Thus:

(iv.α) The species to which Lassie belongs is the same as the species to which Cujo belongs.
(v.α) Lassie and Cujo belong to the species *Canis familiaris*.
(vi.α) Two carnivorous species live in this house.

will be reinterpreted thus:

(iv.β) The conglomerate of all species-mates of Lassie is the same as the conglomerate of all species-mates of Cujo.

(v.β)   Lassie and Cujo are organisms that are parts of the mass of all dog-meat.

(vi.β)   Two conglomerates of conspecific carnivores have parts living in this house.

Crucial for the success of the method is condition (i) that distinct abstracta have distinct concreta representing them, that the conglomerate of all physical entities of one character should never coincide with the conglomerate of all those of another. For this it is more than sufficient that no two distinct physical entities of the relevant sort should overlap, as is the case with biological organisms. Where there is overlap, the simple method may break down. Quine (1950) illustrates the problem in miniature by the case of subfigures of triangular outline and those of square outline in the adjoining figure.

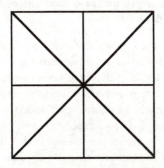

Here the conglomerates of the triangular- and square-bordered regions are the same. It is this difficulty that makes the simple method of elimination inapplicable in the important case of shapes and the related case of expression types.

A less direct and more complicated approach is **contextual** or sentence-by-sentence reduction. In its simplest form, contextual reduction just drops requirement (iii), and allows an X to be represented by more than one x. This makes it impossible to reinterpret a word or phrase naming some one specific X as naming some one specific x, but the idea of the method is precisely that it is not necessary to reinterpret words and phrases in isolation, but only sentences containing them. With this method, any abstracta that can be construed as the characters that concreta of some sort that are equivalent in some sense thereby have in common can be eliminated in favour of those concreta, even if the concreta overlap, and even if conglomerates are not accepted. A species, for instance, can be represented by any organism of that species, *Canis familiaris* by any and every dog, *Felis domestica* by any and every cat. A shape can be represented by any inscribed figure of that shape. (iv.α), (v.α), (vi.α) are then reinterpreted without assuming conglomerates as follows:

(iv.γ) Lassie is a species-mate of Cujo.
(v.γ)   Lassie and Cujo are dogs.
(vi.γ) Two non-conspecific carnivores live in this house.

And moreover what could not be reinterpreted before:

(vii.α) Three convex shapes are inscribed in the above figure.

can now be reinterpreted:

(vii.γ) Three convex figures not like-shaped with each other are inscribed in the above figure.

Even the new approach, however, is only applicable to abstracta that can be construed as characters of actually existing concreta, including the types of expressions that actually have been inscribed or uttered. This presumably meets the needs of a lexicographic theory of words, since such a theory is presumably concerned only with words that actually have been used in a language. It presumably does not meet the needs of a grammatical theory of sentence types, since such a theory is presumably concerned not just with sentences that actually have been used but also with sentences that have not actually been but potentially could be used in the language. In the same way, any attempt to reconstrue talk of numbers in terms of talk of numerals, or talk of sets or properties in terms of talk of predicates defining them, would require infinitely many numerals or predicates. Goodman and Quine declined to assume the infinite extent or divisibility of matter, without which even the acceptance of conglomerates provides only finitely many concreta.

(The notion of contextual representation can be broadened somewhat, by allowing an abstractum X to be represented not by concreta x but rather by ordered k-tuples $x_1, \ldots, x_k$ thereof, for some fixed k. This broader notion of representation is the one to be formally presented in article 2.b. Even with this broader notion, however, finitely many concreta can only represent finitely many abstracta; whereas presumably a grammatical theory of sentences, and certainly a mathematical theory of numbers, assumes infinitely many abstracta.)

Here lies the ultimate source of the failure of Goodman and Quine. Two ways around this obstacle suggest themselves, each involving apparatus they ignored or rejected. One way would be to invoke modal logic and its distinction between actuality and possibility, and reinterpret assertions about sentence-types or numeral-types, not as assertions about what sentence-tokens or numeral-tokens there actually are, but rather as assertions about what tokens there possibly could have been. Another way would assume the infinite extent or divisibility of space, and invoke appropriately shaped regions of empty or heterogeneously occupied space to make up for the lack of regions homogeneously filled with ink (or rather of the contents of such regions, material inscriptions made of ink). Broadly speaking, these two routes are the ones that have been taken by Goodman's and Quine's successors, whose work is to be surveyed in later chapters of this book.

b. Tarskian Reduction

The broadest sense of reduction of (theories about) entities of one sort to (theories about) entities of another sort that is relevant to present concerns seems to have first been explicitly discussed in a monograph (on a topic quite unrelated to nominalism) by Tarski, Mostowski, and Robinson (1953), and hence it might be called **Tarskian** reduction or elimination. It will be well to consider it in some little detail and with some little rigour. Schematically, it works as follows.

Let T in L be a two-sorted theory in a two-sorted language. Let L° be the one-sorted language whose primitives are just the primary primitives of L, and let T° be the one-sorted theory whose axioms are just the primary axioms of T. In jargon, T° in L° may be called the **primary restriction** of T in L: it is the part that is directly about primary entities.

If any result expressible by a closed formula of the language L° of T° that is deducible from T is deducible from T°, then in jargon T is called a **deductively conservative** extension of T°. If for every formula F in the language L of T having only primary free variables there is a formula F° of the language L° of T° with the same free variables such that it is deducible from T that F and F° hold of exactly the same primary entities, then in jargon T is called an **expressively conservative** extension of T°. In this case, the class (if there is just one free variable) or relation (if there are several free variables) determined by a formula F of L is also determined by the formula F° of L°, and in this sense any assertion about classifications of or relationships among primary entities that is expressible in L is (according to T itself) already expressible in $L°$. If T is both expressively and deductively conservative, it is called **fully conservative**. In this case, any information (any notion or result) about primary entities that is supplied by (is expressible in and deducible from) T in L is also supplied by $T°$ in $L°$. Thus if one's interest is only in information about primary entities, all apparatus (all primitives and axioms) pertaining to secondary entities can simply be deleted.

To apply this jargon to the issue of nominalism: standard scientific theory supplies much of the information it supplies about physical entities only indirectly, by way of apparatus pertaining to supposed relationships of physical entities to supposed mathematical entities and supposed classifications of and relationships among the supposed mathematical entities themselves. As much of what science says about observable entities is 'theory-laden', so much of what science says about concrete entities (observable or theoretical) is 'abstraction-laden'. Hence a straightforward formalization of a scientific theory in a two-sorted language would not be fully conservative over its primary restriction. It is in this sense that mathematical entities are present in science and cannot simply be deleted from science.

For instance, elementary chemistry formalized in the most natural way might include the assertion that the atomic weight of beryllium is 9.012182; but there need be no sentence in the theory's primary restriction that comes close to expressing this claim. So a simple deletion of all claims involving secondary entities will say much less than the original theory said about primary entities like beryllium.

Now let T in L be a formalized theory in a formalized language. If an extension T^+ in L^+ of T in L is obtained simply by adding a finite number of new primitives and for each new primitive a single axiom asserting that it holds of exactly the same entities as some old formula, then the new primitives can be regarded simply as abbreviations of the old formulas, and the new axioms simply as their definitions. In this case, T^+ in L^+ is in jargon called a **definitionally redundant** extension of T in L. If a further extension $T^{\dagger}$ of T^+ in the same language $L^{\dagger} = L^+$ is obtained simply by adding finitely many axioms each of which was already deducible (or the instances of finitely many schemes, each of which was already deducible), then the new axioms can be regarded simply as making explicit some implications of the old axioms. In this case, $T^{\dagger}$ in $L^{\dagger}$ is in jargon called an **implicationally redundant** extension of T^+ in L^+, and a **merely redundant** extension of T in L.

A two-sorted theory T in a two-language L will be said to have the **elimination property** if it has a merely redundant extension $T^{\dagger}$ in $L^{\dagger}$ that is fully conservative over its primary restriction $T^{\S} = T^{\dagger \circ}$ in $L^{\S} = L^{\dagger \circ}$. In this case, if one's interest is only in primary entities, then one can take $T^{\S}$ in $L^{\S}$ as a reconstruction of T in L: the new theory eliminates all apparatus pertaining to secondary entities; it retains all information about primary entities supplied by the old theory; and it introduces only novel apparatus that is derivable from (definable from and implied by) the original apparatus, which should be sufficient sanction (explanation and justification) for the acceptability (intelligibility and plausibility) of the novel apparatus from any standpoint from which the original apparatus was proper. Two theories which have a common redundant extension will be called **reformulations** of each other: the same information is derivable from both, and they differ only as to which notions are taken as primitive and which as defined from the primitives, and which assertions are taken as axiomatic and which as deductions from the axioms. (Though they are different sets of sentences and hence different 'theories' as logicians and philosophers tend to use the term 'theory', they are but different formulations of the same 'theory' as mathematicians and scientists tend to use the term 'theory'.)

To apply this jargon to the issue of nominalism: a nominalist would surely welcome a reconstruction or alternative theory that eliminated mathematical entities, and that retained all information about physical entities provided by a standard scientific theory. But there might perhaps arise a question of the philosophical acceptability of the novel apparatus: for the derivability of the novel apparatus from the original apparatus is not in itself sufficient sanction for a nominalist, since the original theory itself was not credited by the nominalist. To obtain a reconstruction or alternative theory whose novel apparatus is philosophically acceptable will be the end or goal of the positive nominalist projects.

Let T in L be a two-sorted theory in a two-sorted language. Let $R(x_1, \ldots, x_k, X)$ be a formula with some positive number k of primary free variables, and a single secondary free variable. The following closed formulas will be called the **existence** and **uniqueness** principles for R:

for any X there is at least one k-tuple $x_1, \ldots, x_k$ such that
$R(x_1, \ldots, x_k, X)$
for any k-tuple $x_1, \ldots, x_k$ there is at most one X such that
$R(x_1, \ldots, x_k, X)$

If these are deducible from T, then R will be said to be a representation formula for T. If there is a **representation** formula for T, then T will be said to have the **representation** property. By modifying the definition of R, arbitrarily designating some one secondary entity (such as o in the case where the secondary entities are real numbers) and arbitrarily stipulating that a k-tuple that did not represent anything else under the original definition of R is to be counted as representing this special secondary entity, one may also assume:

> for any k-tuple $x_1, \ldots, x_k$ there is at least one X such that $R(x_1, \ldots, x_k, X)$

The main metatheorem pertaining to Tarskian elimination states that the representation property implies the elimination property. The proof to be outlined in section 3 provides not only a reconstruction $T^\S$ of T (in the sense that has been made precise), but also a paraphrase or reconstrual § in $L^\S$ of L (in a sense that can also be made precise).

To apply this jargon to the issue of nominalism: for the metatheorem to be applicable, since there are infinitely many mathematical entities to be represented, there will have to be infinitely many physical entities to represent them. If these cannot be supplied by actual material entities, some auxiliary (geometric or modal) apparatus will have to be available in the original theory, and there will arise a question of the scientific accept-ability of the auxiliary apparatus. (In the modal, though not the geomet-ric, case there will arise also the question whether the metatheorem holds for languages and theories based on modal logic as well as for those based on standard logic. This may be considered an aspect of the question of scientific acceptability.) To obtain a representation whose auxiliary appara-tus is scientifically acceptable will be the means or strategy of the posit-ive nominalist projects.

Let there be given, then, a two-sorted theory T in a two-sorted lan-guage L. In the case of most interest, the primary entities will be physical entities and the secondary entities real numbers; then typical primary, mixed, and secondary primitives might be:

(i) x is less massive than y is
(ii) X measures how massive x is
(iii) X is less than Y

Suppose now, as given by the hypothesis of the metatheorem, T in L has the representation property, so that a representation formula R can be defined and the existence and uniqueness principles for R deduced. To show that, as required for the conclusion of the metatheorem, T in L has the elimination property, one needs first to introduce a definitionally redundant extension T^+ in L^+ of T in L; and second to introduce an implicationally redundant extension $T^\ddagger$ in $L^\ddagger = L^+$ of T^+ in L^+; and third to show that $T^\ddagger$ in $L^\ddagger$ is both deductively and expressively conservative over its primary restriction $T^\S = T^{\ddagger\circ}$ in $L^\S = L^{\ddagger\circ}$.

Thus the proof will proceed in three stages. It will be outlined in articles 3.a–3.c in just enough detail to indicate what theory constitutes a reconstruction and what mapping constitutes a reconstrual; further details of the proof that this theory and that this mapping are indeed a reconstruction and a reconstrual will be omitted. (The three spots where the most important omissions occur are marked (*), (**), (***) in the outline; some more details about these will be provided in an optional semi-technical appendix, article 3.e.)

3. METHOD OF ELIMINATION: PROOF OF THE THEOREM

a. *Definitionally Redundant Extension*

To obtain T^+ in L^+, add the following to T in L: for each mixed primitive $G(u, v, \ldots, X, Y, \ldots)$ of L add a **counterpart** primitive $G^\circ(u, v, \ldots, x_1, \ldots, x_k, y_1, \ldots, y_k, \ldots)$, along with the axiom defining it to abbreviate:

$$\exists X \exists Y \ldots (R(x_1, \ldots, x_k, X) \wedge R(y_1, \ldots, y_k, Y) \wedge \ldots \wedge$$
$$G(u, v, \ldots, X, Y, \ldots))$$

For example, the counterpart to (ii) of article 2.b might be read:

(ii°) $x_1, \ldots, x_k$ represent how massive x is

Also, for each secondary primitive $H(X, Y, \ldots)$ of L, add a **counterpart** primitive $H^\circ(x_1, \ldots, x_k, y_1, \ldots, y_k, \ldots)$, along with the axiom defining it to abbreviate:

$$\exists X \exists Y \ldots (R(x_1, \ldots, x_k, X) \wedge R(x_1, \ldots, x_k, Y) \wedge \ldots \wedge$$
$$H(X, Y, \ldots))$$

For example, the counterpart to (iii) of article 2.b might be read:

(iii°) $x_1, \ldots, x_k$ represent less than $y_1, \ldots, y_k$ do *b. Implicationally Redundant Extension*

To each formula $Q(u, v, \ldots, X, Y, \ldots)$ of L^+, associate a **counterpart** formula $Q^\circ(u, v, \ldots, x_1, \ldots, x_k, y_1, \ldots, y_k, \ldots)$ by making the replacements shown in the adjoining table.

Counterpart Formulas

Replace each	By
secondary variable X	a k-tuple of new primary variables $x_1, \ldots, x_k$
secondary primitive H	its counterpart primitive H°
mixed primitive G	its counterpart primitive G°
secondary quantification $\forall X$	the k-fold primary quantification $\forall x_1 \ldots \forall x_k$
secondary quantification $\exists X$	the k-fold primary quantification $\exists x_1 \ldots \exists x_k$

To obtain $T^\dagger$ in $L^\dagger = L^+$, add the **counterpart** axiom B° of each mixed axiom B of T^+. Add also the **counterpart** axiom C° of each secondary axiom C of T^+. It can be shown that (*) these counterpart axioms are already deducible from T^+, so that $T^\dagger$ is implicationally redundant over T^+.

c. Primary Restriction

Consider the primary restriction $T^\S = T^{\dagger\circ}$ in $L^\S = L^{\dagger\circ}$ of $T^\dagger$ in $L^\dagger$. It can be shown that (**) if Q is a formula of $L^\dagger$ with only primary free variables, then it is deducible from $T^\dagger$ that Q and Q° hold of exactly the same objects, so that $T^\dagger$ is also expressively conservative over $T^\S$. It can also be shown that (***) if a closed formula Q of $L^\S$ is deducible from $T^\dagger$ then Q is deducible from $T^\S$, so that $T^\dagger$ is deductively conservative over $T^\S$, and hence fully conservative over $T^\S$. Thus $T^\S$ in $L^\S$ is a reconstruction of T in L in a sense that has now been made precise. The restriction § of the mapping $^\circ$ of formulas of $L^\dagger$ to formulas of $L^\S$ to a mapping of formulas of L to formulas of $L^\S$ is a reconstrual of L in a sense that can be made precise.

d. Recapitulation

It may be desirable to summarize just as much of the foregoing discussion as will need to be remembered in Parts II and III. The input or starting-point for each of the various sample or specimen nominalist projects to be presented there will be a (formalized version of a) standard scientific theory, a two-sorted theory T in a two-sorted language L, with primary variables x, y, z, ... for physical entities and secondary variables X, Y, Z ... for real numbers. The language L will have some primary primitives F, some mixed primitives G, and it will have as the secondary primitives H those of analysis. For purposes of illustration, the following may serve as paradigms of such F, G, and H:

(i) x is less massive than y is
(ii) X measures how massive x is
(iii) X is less than Y is

The theory T may have some primary axioms A, will have some mixed axioms B, and will have as secondary axioms C those of analysis. Such a theory may be called **analytically formulated**.

In each strategy, L and T will also be assumed to include either geometric or else modal apparatus, depending on the strategy, either including geometric along with material entities among the physical entities, or including modal operators along with elementary operators in the logic. The scientific status of this auxiliary apparatus will have to be considered. The overall method for the sample or specimen nominalistic strategies to be presented in Part II and some beyond will be to define a notion 'x_1, ..., x_k represent X' and deduce the existence (actual or possible) for any real number of physical entities (material or geometric) representing it, and the uniqueness of the real number represented by any (actual or possible) physical entities (material or geometric). This done, a reconstruction in a one-sorted theory $T^\S$ of T and a reconstrual [§] mapping L to its one-sorted language $L^\S$ can be obtained by what has been called Tarskian elimination. These will be the outputs or stopping-points of the nominalist project.

The language $L^\S$ will have as primary primitives F° the original primary primitives F of L, as mixed primitives G° the counterparts of the original mixed primitives G of L, and as secondary primitives H° the counterparts of the original secondary primitives H of analysis. Paradigmatically:

(i°) x is less massive than y is
(ii°) x_1, ..., x_k represent how massive x is
(iii°) x_1, ..., x_k represent less than y_1, ..., y_k do

The theory $T^\S$ will have as primary axioms A° the original primary axioms A of T, as mixed axioms the counterparts $B^\S$ of the mixed axioms of T, and as secondary axioms $C^§$ the counterparts of the axioms C of analysis.

The philosophical status of this novel apparatus will have to be considered. The issues of the scientific and philosophical status of the auxiliary and the novel apparatus cannot be settled once for all at the level of generality of the present discussion, and will have to be considered case by case. (In the modal case, an aspect of these otherwise philosophical and intuitive issues will be the logical and technical issue whether the reconstrual $§$ preserves logical deducibility as it ought.) This, then, is the framework that will be adopted throughout later chapters.

e. Some Details of the Proof

The sense in which $§$ is a reconstrual is that, first, it leaves primary formulas unchanged, and second, it leaves logical connections of deducibility unchanged. The first fact follows immediately from the definition of $°$, and the second from the fact that $°$ leaves logical structure (negation, conjunction, disjunction, universal and existential quantification) unchanged, which fact itself follows immediately from the definition of $°$. This fills the gap at the spot marked (***) in the foregoing outline. (The question whether the metatheorem holds for modal as well as for elementary logic amounts to the question whether $°$ preserves deducibility in this sense in a modal as well as in an elementary context.)

To fill the gaps at the spots marked (*) and (**) it would suffice to establish that the following is deducible from T^+ for any Q:

(#)   for any u, v, ... and any X, Y, ... and any
$$x_1, \ldots, x_k, y_1, \ldots, y_k, \ldots$$
if $R(x_1, \ldots, x_k, X)$ and $R(y_1, \ldots, y_k, Y)$ and ..., then
$Q(u, v, \ldots, X, Y, \ldots)$ if and only if
$Q°(u, v, \ldots, x_1, \ldots, x_k, y_1, \ldots, y_k, \ldots)$

Note (**) requires only the special case where there are no free secondary variables, and (*) requires only the very special case where there are no free secondary or primary variables. Here (#) can be established by induction on the logical complexity of Q. For the base step (atomic formulas), in the non-trivial case (Q a mixed primitive G or secondary primitive H), use the defining axioms (for $G°$ or $H°$). For the induction step (compound formulas), in the non-trivial case (Q a universal or existential quantification with respect to a secondary variable), use the existence and uniqueness principles (for the representation notion in question).

4. NON-TARSKIAN REDUCTION

a. *Skolemite Reduction*

Two alternative methods of reduction that promise to be less work—but that deliver what are generally considered less satisfactory results—may be briefly mentioned as foils to Tarskian reduction. Like Tarskian reduction they are named, not for philosophical advocates of their use for nominalistic purposes, but rather for logicians whose technical work they exploit. If one drops the demand that the representation notion be definable from the primitives of a given language L and that the representation assumptions of existence and uniqueness be deducible from the axioms of the given theory T, then the easier method of **Skolemite** reduction becomes applicable, at least if T implies the existence of infinitely many primary entities (and to avoid trivialities, of infinitely many secondary entities). On this method, one simply adds a new predicate $x \circledR X$ for representation, to give a new language $L^\dagger$, and simply assumes existence and uniqueness in a very strong form:

($\ddagger$)   for every X there exists a unique x such that $x \circledR X$ and
         for every x there exists a unique X such that $X \circledR x$

as new axioms, to give a new theory $T^\dagger$. One then proceeds as in Tarskian reduction. Though there can be no question of expressive conservativeness with this method, $T^\dagger$ can be proved deductively conservative over T. The proof of this result may be briefly outlined.

The **Completeness Theorem** of Kurt Gödel connects **syntax** in the logicians' sense of **proof theory**, with **semantics** in the logicians' sense of **model theory**. The central notion of the former is that of the **deducibility** of a formula R from a theory T, defined to mean that there exists a finite sequence of formulas constituting a **deduction** of R from T. The central notion of the latter is that of a formula R being a **consequence** of a theory T, defined to mean that R is true in all **models** in which (every axiom of) T is true. The theorem states that the two notions coincide.

Here a model for a language L consists of a universe Γ and a specification for each primitive $F(x, y, \ldots)$ of L and each $a, b, \ldots$ in Γ of whether or not $F(a, b, \ldots)$ is to count as true. Then truth is defined for molecular formulas as in the adjoining table (with the cases of $\wedge$ and $\forall$ being analogous to those for $\vee$ and $\exists$). The last line of the table is needed only for a two–sorted language, where there must be a second universe Δ.

Model Theory for Standard Logical Apparatus

Formula	Truth condition
$\sim Q(a, b, \ldots)$	$Q(a, b, \ldots)$ is not true
$Q(a, b, \ldots) \vee R(a, b, \ldots)$	$Q(a, b, \ldots)$ is true or $R(a, b, \ldots)$ is true
$\exists x\, Q(a, b, \ldots, x)$	for some c in Γ, $Q(a, b, \ldots, c)$ is true
$\exists X\, Q(a, b, \ldots, X)$	for some D in Δ, $\exists x\, Q(a, b, \ldots, D)$ is true

Using the Completeness Theorem, the result stated above is equivalent to the following: if T implies the existence of infinitely many primary entities (and infinitely many secondary entities) and $L^\ddagger$ is the result of adding a new primitive ® to L and $T^\ddagger$ the result of adding (‡) to T, then for any formula R of L that is not a consequence of T, this R is still not a consequence of $T^\ddagger$: if there exists a model $\mathcal{M}$ of T in which R is not true, then there exists a model $\mathcal{N}$ of $T^\ddagger$ in which R is not true. This result can be proved using the Transfer Theorem of Leopold Löwenheim and Thoralf Skolem, according to which, if there is a model of T in which R is not true, then there is such a model $\mathcal{M}$ whose universe Γ is (or in the present, two-sorted case, whose universes Γ and Δ are) countable (or finite, an alternative ruled out in this case by assumption). Since the two universes are of the same, countable, cardinality, there is a bijective relation ρ between the elements of the one and those of the other. By interpreting a®b to be true if and only if a is ρ-related to b, one obtains a model $\mathcal{N}$ for the larger language $L^\ddagger$ and moreover one in which the larger theory $T^\ddagger$ is true, as required.

There seems to be a consensus among nominalists engaged in positive programmes that Skolemite eliminations are unacceptable. Perhaps the thought is that the ideological costs involved in accepting a completely unexplained representation primitive and completely unjustified existence and uniqueness axioms outweigh the ontological benefits: that the Skolemite theory leaves one almost as far from any nominalistic explanation of the observed past success of standard scientific theories in making predictions about concreta or any nominalistic justification as does the instrumentalist 'theory' consisting of the bare assertion that concreta behave as if abstracta existed and standard scientific theories were true.

b. Craigian Reduction

If one drops the demand for a language and theory with only finitely many primitives and finitely many axioms (or schemes), then the trivial method of **Craigian** elimination is applicable. On this method, given a theory T in a language L, one just adds for every formula P of the language L with no free secondary variables a new primitive F_P, along with an axiom B_P defining F_P to hold of exactly the same primary entities as did P, to obtain an extended theory T^+ in an extended language L^+. And one can just add every formula A with no secondary variables of this extended language that is deducible from the axioms of the extended theory to form a further extended theory $T^{\dagger}$ in the same extended language $L = L^+$. One can then discard from the extended language and theory $L^{\dagger}$ and $T^{\dagger}$ thus obtained all primitives and axioms pertaining to secondary variables, to obtain a language and theory L° and T°.

The language L° here has infinitely many primitives, though each can still be taken to be a finite sequence from a finite alphabet, and it is **effectively decidable** whether or not a finite sequence from that finite alphabet does constitute such a primitive: there is a mechanical procedure for determining this in every case. The theory T^+ has infinitely many axioms, but it is also effectively decidable whether or not a given formula of its language is an axiom. The theory T° here has infinitely many axioms, and further it is not effectively decidable whether a given formula A of its language L° (that is not already an axiom of T^+) is supposed to count as an axiom or not, since it will be so if and only if there is a deduction of it from T^+, and while there is a mechanical procedure for determining whether a given finite sequence of formulas constitutes such a deduction, there is not in general a mechanical procedure for determining for a given formula A whether there exists such a deduction, by the **Undecidability Theorem** of Alonzo Church. However, there is a theory T^∞ from which the same formulas are deducible as from T° which is thus effectively decidable, according to the **Reaxiomatization Theorem** of William Craig. The proof of this result may be briefly outlined.

The proof of reaxiomatization uses the fact that there are only countably many finite sequences from any finite alphabet, so that natural numbers can be assigned as codes to deductions, in such a way that one can go back and forth between code number and finite sequence in an effective or mechanical fashion. The theory T° was to have as axioms all those formulas A for which there is a deduction of the appropriate kind. The theory T^∞ has as axioms all those formulas B such that B is the conjunction $(A \wedge \ldots \wedge A)$ of some number n of copies of some formula A with itself, and such that n is the code number of deduction of A of the appropriate kind.

Discussion of the Craigian method in the literature sometimes omits the preliminary step of enlarging the language from L to L^+. So presented, the method invites the objection that the Craigian alternative to a theory is empirically impoverished or causally non-explanatory. And indeed, if the original theory has only a handful of primary predicates, its primary implications will hardly capture all the empirical or causal information about concrete entities contained in the original theory. (Recall the hydrogen sulphide example in section I.A.3 and the beryllium example in article 2.b.) So neither will any reaxiomatization of its primary implications. This objection—a fairly standard one in the literature—does not apply to the version of the Craigian method presented here.

Nonetheless, there seems to be a consensus among nominalists engaged in positive programmes that Craigian eliminations (even of the kind presented here) are unacceptable. Perhaps the thought is that the Craigian 'theory' is little more than a formal counterpart of the instrumentalist 'theory' consisting of the bare assertion that concreta behave as if abstracta existed and standard scientific theories were true. Ultimately, the grounds for dissatisfaction with non-Tarskian reductions depend on the grounds for dissatisfaction with instrumentalism, on the grounds for dissatisfaction with a merely negative, destructive nominalism, on the motivation for engaging in a positive, reconstructive nominalistic project in the first place.

PART II

Three Major Strategies

A

A Geometric Strategy

o. OVERVIEW

The geometric strategy of nominalistic reconstruction faces two main tasks, one technical, one philosophical. On the technical side, geometric nominalism seeks to eliminate numerical entities in favour of geometric entities. In traditional mathematical as opposed to philosophical usage, pure geometry in the Euclidean style is called **synthetic**, and coordinate geometry in the Cartesian style is called **analytic**. The technical task may thus be described as that of producing synthetic alternatives to standard analytic formulations of scientific theories. On the philosophical side, geometric nominalism seeks to persuade nominalists to be indulgent towards geometricalia, to admit them as concrete. In contemporary philosophical usage, **substantival** and **relational** views of space are understood as the acceptance and the rejection of such geometric entities as points and regions of space. The philosophical task may thus be described as that of defending substantivalism against relationalism against a background of nominalism. On both sides, geometric nominalism has deep historical roots in the work of early modern physicists and geometers, which can only be briefly noted in the present chapter. (For more information, the reader is referred to the standard reference work Kline (1972) on the history of mathematics and the basic survey Sklar (1974) of the philosophy of space and time.) It also has extensive intellectual debts to more recent logicians and philosophers.

Presumably the alternatives to standard formulations of scientific theories that geometric nominalism seeks to provide should be more 'elegant' or 'attractive' than the instrumentalist 'theory' consisting of the bare assertion that concreta behave 'as if' abstracta existed and standard scientific theories were true. Just how much more 'elegant' the alternatives must be depends on just what the grounds for dissatisfaction with the 'as if' theory are supposed to be. Surely the theories for which geometric nominalism seeks to provide alternatives to the standard formulations should be 'realistic', not in any philosophical sense, but in the everyday sense of being based on the most up-to-date science. For a philosophical thesis should not be based on a scientific falsehood. The main dilemma for geometrical nominalism is this: the case for accepting geometric entities as concrete draws on realistic, contemporary, twentieth-century physics; but the most elegant elimination of numerical entities in favour of such geometric entities can be carried out only for unrealistic, classical, nineteenth-century physics. It remains an open question how attractive a nominalistic alternative to up-to-date physics can be developed.

This chapter provides a brief discussion of background on syntheticism and substantivalism in section 1. An elegant treatment of unrealistic physics is exhibited in some detail in sections 2–4, both in order to indicate the kind of thing one might hope some day to achieve for more realistic physics, and in order to provide background to a discussion in article 5.a of the difficulties standing in the way of such an achievement at present, and in the optional semi-technical appendix article 5.b of a conceivable further difficulty that might arise in the future.

1. HISTORICAL BACKGROUND

Throughout the early modern period it was the consensus view that the basic objects of algebra, real numbers, are to be identified with ratios of lengths, areas, or volumes, and that the basic operations and laws of algebra are to be explained and justified in terms of geometric constructions and theorems. Even the pioneers who introduced coordinate methods, and thereby made the techniques of algebra and calculus available for application to problems of geometry and mechanics, still regarded the older pure geometry as providing the foundation for algebra. Such an attitude is hinted at, somewhat obscurely and confusingly, in the opening paragraphs of René Descartes's *Géométrie*, and is expounded, clearly and distinctly, in the opening pages of Isaac Newton's *Universal Arithmetick*. Contemporary geometric nominalism goes further than the early modern consensus by requiring that analytic methods, however useful in the context of discovery, are not to be mentioned in the context of justification. But even for this there is precedent in Newton's practice in his *Principia*, though his example was not imitated by many of his successors, such as Pierre Simon de Laplace in the *Mécanique Céleste*. Present-day geometric nominalism, of course, would wish to conform to twentieth-century standards of rigour, which are higher than were those of the seventeenth century. A more rigorous synthetic geometry and geometric algebra than that of Descartes or Newton is provided by the famous monograph of their successor David Hilbert (Hilbert 1900). Further logical refinements can be found in the work of Alfred Tarski and his school, beginning with Tarski (1959).

Long before the time of Hilbert and Tarski, however, the foundational significance of the reduction of algebra to geometry had come to seem doubtful. For the discovery of non-Euclidean geometry in the early 1800s undermined the early modern consensus according to which geometry was a foundation for algebra and analysis, doubts being expressed already by C. F. Gauss, one of the main co-discoverers of the new geometries. Already with Gauss, and more explicitly and emphatically with Bernhard Riemann, one finds the view, which has become the later modern consensus, that one must distinguish **mathematical geometry** from **physical geometry**. One should perhaps speak rather of mathematical 'geometries' in the plural, for they are legion. These myriad geometries are not conflicting opinions about some one and the same object, space, but rather are definitions of various different classes of mathematical 'spaces'. If one does speak after all of mathematical 'geometry' in the singular, it must be understood as the comparative studies of all these different 'spaces'. (These mathematical 'spaces' include, besides those mathematical structures, such as 'Riemannian manifolds', that have at one time or another been proposed as images of physical space, many others, namely, all those, such as 'Hilbert space', that are sufficiently similar to allow them to be fruitfully investigated by similar methods, regardless of what the nature of the applications if any of these other structures may be.)

Inverting the earlier order of things, the 'points' of these 'spaces' are now standardly taken to be set-theoretically generated out of real numbers, so that algebra and analysis become the foundation rather than the superstructure. In effect, a 'point' of a k-dimensional mathematical 'space' is often simply identified with the k-tuple real numbers that are its coordinates. This inversion was an important motive for the search for a new, non-geometric foundation for the numerical side of mathematics, now standardly taken to be provided by the set-theoretic generation of the real numbers out of the natural numbers, themselves set-theoretically generated (a process briefly outlined in article I.B.1.a). In so far as it seeks to restore geometry as a foundation for mathematics, geometric nominalism is not revolutionary but counter-revolutionary.

According to the later modern consensus, mathematicians collect various different mathematical 'spaces', while physicists select a single one from among them as an image of physical space, the selection being made on empirical grounds. However, physical geometry, by itself and without auxiliary hypotheses, makes no empirical predictions. For instance, an attempt to test the Euclidean vs. non-Euclidean hypotheses about the sum of angles in a triangle by surveying some large triangle on earth or in the heavens, as was done by Gauss, never tests just these geometric hypotheses alone. For even the use of the most low-tech surveying instruments, the sextant and the plumb, involves auxiliary hypotheses about light and weight: the hypotheses that light travels, and weights fall, in straight lines. Ultimately the selection of a physical geometry is inseparable from the choice of an electromagnetic theory and of a gravitational theory, a point much emphasized by Henri Poincaré. Thus while mathematical geometry becomes a branch of the theory of sets, physical geometry becomes a component of the theory of electromagnetism and gravitation.

But one perhaps should not use the term 'geometry'—or any of the terms in the spatial or temporal column in the adjoining table. For whether it is acceptable to speak in physics of spatial 'points' and so forth depends on whether it is acceptable to speak of absolute rest, to speak of being at the same place on different occasions.

Terminology

Spatial	Temporal	Combined	Neutral
geometry	chronometry	geometry-chronometry	kinematics
points	instants	point-instants	events
space	time	space-time	world

Likewise, whether it is acceptable to speak of 'instants' and the like depends on whether it is acceptable to speak of absolute simultaneity, to speak of being at different places on the same occasion.

Now though **absolutism** vs. **relativism** was a live issue as regards rest around 1700, and as regards simultaneity was still a live issue around 1900, the victory of relativism since the work of Albert Einstein has been complete. On further thought, perhaps precisely because the defeat of absolutism has been so very complete there will no longer be any serious danger of confusion if one after all uses 'geometry' and other spatial and temporal terms, understanding them as colloquial abbreviations for the more cumbersome but more accurate expressions in the combined or neutral columns of the table.

Contemporary geometric nominalism is committed to substantivalism as opposed to relationalism, to the acceptance as opposed to the rejection of such geometric entities as space, regions, and points. Centuries ago, when Newton and G. W. von Leibniz were debating these issues, substantivalism and relationalism were tightly intertwined with absolutism and relativism, respectively. Fortunately for geometric nominalism, in the course of the long history of such debates the two issues became disentangled. Indeed, present-day substantivalist arguments, of which the best-known is perhaps that in the position paper Earman (1970), are based specifically on relativistic considerations.

One consequence of the kind of division of labour instanced by the distinction between mathematical and physical geometry is that it permits mathematics to progress by addition, while physics has to progress by amendment. And there have been, since the nineteenth century, several successive amendments to the Newtonian physical geometry. First, there was **non-relativistic** geometry, implicit in classical gravitational theory. Second, there was and is **special-relativistic** geometry, implicit in classical electromagnetic theory—though it took quite a bit of work to make it explicit—and retained by the quantum theories that provide the best currently available accounts of electromagnetic and weak and strong nuclear forces. Third, there is **general-relativistic** geometry, explicit in the best currently available theory of gravitation. Fourth, since the best currently available theories of electromagnetic and nuclear phenomena on the one hand and of gravitational phenomena on the other hand are each presumed to require amendment in order to take account of the other, there will eventually have to be a **super-relativistic** geometry in the much hoped-for 'final theory of everything'. At present there is no consensus even as to how many dimensions the geometry of a theory unifying quantum mechanics and general relativity would ascribe to physical space. (Figures as high as 26 have been mentioned.)

Though the occurrence of amendments has seemed to many to argue, if not in favour of founding mathematics on set theory, at least against syntheticism, against founding mathematics on geometry, still the content of the amendments has seemed to many to argue in favour of substantivalism, in favour of regarding geometric-chronometric entities as physical. One main consideration arises already in connection with the shift from the matter-theoretic standpoint of classical, Newtonian gravitational theory to the field-theoretic standpoint of classical, Maxwellian electromagnetic theory. In the latter, charged bodies do not act instantaneously at a distance by electromagnetism on other charged bodies, as in the former massive bodies act instantaneously at a distance by gravity on other massive bodies. Rather, electromagnetic action propagates at a finite speed, that of light, and as a consequence, much of the energy of a physical system cannot be localized in material bodies, but must rather be ascribed to a force-field between them extending throughout space. (When the theory is formulated special-relativistically, the very distinction between mass and energy blurs.) According to many philosophical commentators, the force-field must be considered to be a physical entity, and as the distinction between space and the force-field may be considered to be merely verbal, space itself may be considered to be a physical entity.

The other main consideration arises in connection with the shift to general-relativistic, Einsteinian gravitational theory. This blurs or abolishes, as regards gravity, the distinction between kinematics and dynamics, or between inertial and accelerated motion. The presence of massive bodies is not taken to exert a gravitational force deflecting the motion of other bodies from the straight path in which they would otherwise move. Rather, it is taken to deform space itself, making the straightest paths available for a body to move in more curved than they would have been in the absence of those massive bodies. According to many philosophical commentators, space may hence be considered a causal patient and agent, shaped by the presence of massive bodies, and constraining the motion of other bodies by the paths it makes available. However reactionary it may be in seeking to base algebra and analysis on physical geometry, geometric nominalism is thoroughly progressive in its substantivalism, in so far as that substantivalism is based on such considerations as the foregoing. It must be recognized, however, that just how much support such considerations can provide for the acceptance of geometricalia by nominalists depends in part on just what is supposed to be the objection to abstracta, the motivation for nominalism, in the first place.

2. TECHNICAL BACKGROUND

At a very fundamental level of mathematical geometry comes **affine basic coordinate plane geometry**. Here **plane** (as opposed to **solid** or **hyperspace**) of course indicates a geometry of two (as opposed to three or four) dimensions, while **coordinate** (as opposed to **pure**) indicates an analytic theory involving both geometrical and numerical entities rather than a synthetic theory involving only the former and avoiding the latter. Here also **basic** (as opposed to **intermediate** or **higher**) indicates a theory in which the geometric entities are points (as opposed to special regions or arbitrary regions), while **affine** indicates a kind of minimal theory common to Euclidean and several non-Euclidean geometries.

The geometry with the above long name may be formalized as a two-sorted theory $T^\dagger$ in a two-sorted language $L^\dagger$. The language $L^\dagger$ will have variables $x, y, z, \ldots$ for plane points, and variables $X, Y, Z, \ldots$ for real numbers. It will have one primary primitive, for the **order** relation among points, 'y lies between x and z'. It will have the three secondary primitives of analysis for the order, sum, and product relations among real numbers, but not for integrity. And it will have two mixed primitives for (**preferred**) **coordinates**:

> X is the horizontal coordinate of x
> (on some preferred coordinate system)
> X is the vertical coordinate of x
> (on this preferred coordinate system)

The theory $T^\dagger$ will have no primary axioms. It will have the secondary axioms of analysis (in the sense of article I.B.1.b) except those for integrity, the basic algebraic axioms for order, sum, and product, and the continuity scheme. It will have two mixed axioms of (**preferential**) **coordination**:

(i)   for every point x there exists a unique pair of real numbers X_1, X_2
        such that
        X_1, X_2 are the coordinates of x (on the preferred coordinate system),
        and
        for every pair of real numbers X_1, X_2 there exists a unique point
        x such that
        X_1, X_2 are the coordinates of x (on the preferred coordinate system)

(ii)   for any points x, y, z with coordinates $X_1, X_2, Y_1, Y_2, Z_1, Z_2$
        (on the preferred coordinate system),
        y lies between x and z if and only if
        there exists a real number U such that $0 \le U \le 1$ and
        $Y_1 = X_1 \cdot U + Z_1 \cdot (1 - U)$ and $Y_2 = X_2 \cdot U + Z_2 \cdot (1 - U)$

(Here the usual symbols for addition, multiplication, and so forth have been used. The second axiom can be, and officially should be, written out using just the order, sum, and product predicates.)

In older formalizations of geometry, as in Hilbert's famous monograph already cited, other geometric entities beyond points, beginning with straight lines, are mentioned. There are variables $\xi, \upsilon, \zeta, \ldots$ for lines, and primitives for relationships between points and lines or among lines, such as **incidence**, 'x lies on ξ' or 'ξ goes through x', and **parallelism**, 'ξ goes parallel to υ'. But these are in a sense superfluous, since lines can be represented by or reduced to points, and since the relationships among points that are the counterparts of the incidence and parallelism relationships above, namely, the **collinearity** relation and the **equidirectedness** relation:

> x lies aligned with x_1 and x_2
>
> x_2 lies with respect to x_1 parallel to how y_2 lies with respect to y_1

are already definable in terms of order, by the conditions:

> x lies between x_1 and x_2, or x_1 lies between x_2 and x, or
> x_2 lies between x and x_1
>
> there exists no z such that z is collinear both with x_1 and x_2 and with y_1 and y_2

A semi-popular exposition of formalized geometry (Tarski 1959) puts it as follows:

Thus, in our formalization . . . only points are treated as individuals. . . . [Our] formalization does not provide for variables of higher orders and no symbols are available to . . . denote geometrical figures. . . . It should be clear that, nevertheless, we are able to express in our symbolism . . . results which can be found in textbooks and which are formulated there in terms referring to special classes of geometrical figures such as the straight lines . . . the segments, the triangles, the quadrilaterals, and more generally the polygons with a fixed number of vertices . . . This is primarily a consequence of the fact that, in each of the classes just mentioned, every geometrical figure is determined by a fixed finite number of points. For instance, [in the notation of the present work] instead of saying that a point z lies on the straight line through the points x and y, we can state that either x lies between y and z or y lies between z and x or z lies between x and y . . .

Using these notions, several important assertions of elementary geometry can be expressed. Notable among these is the version of Euclid's parallel postulate assertion known as **Playfair's Postulate**:

> for any x, y, z that are not collinear, there exists a point u
> such that no point is collinear both with x and y and with z
> and u;
> and if v is any other such point, then z is collinear with u and v

Also, the assertion known as **Desargues's Theorem** and illustrated in the adjoining figure:

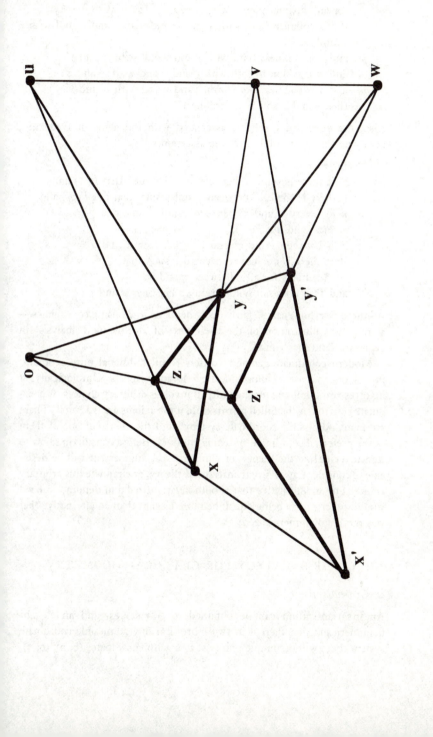

for any distinct points o, x, y, z, x', y', z', u, v, w,
if o is collinear both with x and x' and with y and y' and with z
and z',
and v is collinear both with x and y and with x' and y',
and w is collinear both with y and z and with y' and z',
and u is collinear both with z and x and with z' and x',
then u and v and w are collinear

Also, the **continuity** scheme associated with the name of Dedekind, involving for any formula $P(z)$ the assertion:

for every x and y
if there is some z' between x and y such that $P(z')$ and
if there is some z'' between x and y such that not $P(z'')$ and
if for every z' and z'' between x and y such that
$P(z')$ and not $P(z'')$,
z' is between x and z'' and z'' is between z' and y
then there is a w between x and y such that
$P(z)$ holds for all z between x and w,
and $P(z)$ does not hold for any z between w and y

All these assertions can be fairly easily deduced in T^+ using coordinates—a miniature illustration of the usefulness of algebra (and analysis) in geometry (and mechanics).

Modern coordinate geometry differs from traditional pure geometry in not one but two philosophically relevant respects. Ontologically, it involves real numbers; ideologically, it involves arbitrary choices. A coordinate system may be called **admissible** if the axioms are true on it. There are many admissible coordinate systems, and the choice of one of them as the preferred coordinate system is arbitrary. Before considering how to **denumericalize** the theory, or eliminate real numbers, it will be desirable to consider how to **invariantize** the theory, or eliminate this arbitrary choice. For an alternative that is both invariantized and denumericalized would, other things being equal, be more elegant than an alternative that was merely denumericalized.

3. THE STRATEGY FOR CLASSICAL GEOMETRY

a. *Generalization*

An invariantization can be obtained in two stages, and an elegant denumericalization thereof in two more. For any admissible coordinate system there will be unique points u, v, w with coordinates $(0, 0)$, $(0, 1)$, and (1, 0) on that system, by the coordination axiom (i) of section 2. These may be called the **benchmarks** of the coordinate system. (For collinear points u, v, w there will be no admissible coordinate system having them as benchmarks, by the coordination axiom (ii) of section 2.) It can be shown that for non-collinear points u, v, w there will be a unique admissible coordinate system having them as benchmarks. The coordinates on this alternative system of any point x can be obtained algebraically, by what is called an 'affine transformation' and is explained in any good textbook on linear algebra, from the coordinates on the preferred system of the point x and of the points u, v, w.

One thus obtains a definitionally redundant extension $T^{\#}$ in $L^{\#}$ if one adds to $T^{\dagger}$ in $L^{\dagger}$ two new primitives for the **generalized coordinate** notions:

> [u, v, w are not collinear and]
> X is the horizontal coordinate of x with respect to
> (the coordinate system having as benchmarks) u, v, w

> [u, v, w are not collinear and]
> X is the vertical coordinate of x with respect to
> (the coordinate system having as benchmarks) u, v, w

along with a defining axiom (iii) indicating how the generalized coordinates with respect to u, v, w of a point x are obtained algebraically from the preferred coordinates of the point x and of the points u, v, w. To any formula Q of $L^{\#}$ one may associate a generalization Q^{+}, obtained as follows: first prefix the formula by a universal quantification, 'for any non-collinear u, v, w ...', and then replace each occurrence of '... preferred coordinates ...' by '... generalized coordinates with respect to u, v, w ...'. It can be shown that the generalizations (i^{+}), (ii^{+}), (iii^{+}) of the axioms (i), (ii), (iii) are obtainable algebraically from the latter axioms. One thus obtains an implicationally redundant extension T^{l} in $L^{\mathsf{l}} = L^{\#}$ if one adds to $T^{\#}$ in $L^{\#}$ the former axioms, along with the trivial axiom:

(0^{+})   there exist non-collinear points u, v, w

T^{l} in L^{l} is a merely redundant extension of $T^{\dagger}$ in $L^{\dagger}$.

b. *Invariantization*

Consider now the restriction T^{+} in L^{+} of T^{l} in L^{l} obtained by deleting the primitives for preferred coordinates and the axioms involving them. It is trivially seen that every formula of L^{+} is invariant, or independent of arbitrary choices, whereas it is equally trivially seen that not every formula of $L^\dagger$ is invariant. Hence $T^\dagger$ in $L^\dagger$ will not be an expressively conservative extension of T^+ in L^+. However, it can be shown that any formula Q of $L^\dagger$ that is invariant is equivalent to its generalization Q^+, which is a formula of L^+. Hence $T^\dagger$ in $L^\dagger$ will be what may be called an 'invariantly expressively conservative' extension of T^+ in L^+. It can also easily be shown that generalization carries axioms of $T^\dagger$ to axioms of T^+, preserves logical relations of implication, and leaves formulas of L^+ unchanged. Hence $T^\dagger$ in $L^\dagger$ is a deductively conservative extension of T^+ in L^+. Thus T^+ in L^+ captures the invariant content of $T^\dagger$ in $L^\dagger$ and hence of the original $T^\dagger$ in $L^\dagger$. It remains to obtain an elegant $T^\S$ in $L^\S$ that will capture the non-numerical content of T^+ in L^+.

c. Denumericalization

The strategy now is to apply the general method of Chapter I.B. To do so one must be able to define a notion:

> the k-tuple of points $x_1, \ldots, x_k$ represents the real number X

and deduce the principles of uniqueness and existence:

> every k-tuple of points represents at most one real number
> every real number is represented by at least one k-tuple of points

This can be done by drawing on the traditional conception of real numbers as ratios of lengths. More formally, the required representation notion amounts to:

> x_0, x_1, x represent X if and only if x is collinear with x_0 and x_1, and X is the ratio of the distance from x_0 to x to the distance from x_0 to x_1, (taken with a positive sign if x_1 is between x_0 and x or x is between x_0 and x_1 and with a negative sign if x_0 is between x and x_1)

which can be expressed algebraically in terms of the coordinates X'_0, X''_0, X'_1, X''_1, X', X'' of x_0, x_1, x with respect to some/any admissible u, v, w, as follows:

$$X' = X'_0 + X \cdot X'_1 \text{ and } X'' = X''_0 + X \cdot X''_1$$

Applying the method of Chapter I.B, one obtains a denumericalization $T^\pm$ in $L^\pm$, the primary restriction of a merely redundant extension of T^+ in L^+; but it is inelegant in two respects.

First, $L^\pm$ involves several new primary primitives. These include a primitive for the counterpart, which may be called the **subproportionality**, to the order primitive on numbers. It amounts to:

> the ratio of the distance from x_0 to x to the distance from x_0 to x_1
> is less than
> the ratio of the distance from y_0 to y to the distance from y_0 to y_1

Similarly for the other algebraic primitives and the generalized coordinate primitives. Second, $T^\pm$ involves several new primary axioms. These are the counterparts of the basic algebraic axioms, of the continuity scheme for numbers, and of the generalized coordinate axioms as mentioned under article 3.b above. Except for the counterpart of the continuity scheme for numbers, which amounts to something very like the continuity scheme for points, these axioms have an artificial look from a geometric viewpoint, though they are counterparts of axioms that had a natural look from an algebraic viewpoint. Thus the ontological benefit (from a nominalist viewpoint) of eliminating numbers is accompanied by the ideological costs of artificial new primitives and artificial new axioms.

d. Beautification

But the long tradition of geometric algebra, beginning with the later books of Euclid's Elements, continuing through the medieval Arabs and on to Descartes, Newton, and their contemporaries, further advanced in the nineteenth century to a culmination in Hilbert, now explained in any good textbook on the foundations of geometry, with final logical refinements by Tarski and his school, can be drawn upon to reduce these costs. First, it can be shown that no new primitives are needed: all the new primitives of $L^\pm$ are definable using elementary logic from the single primitive of order for points. Notably, subproportionality, mentioned under article 3.c above, can easily be defined in terms of **proportionality**:

$x_0x : x_0x_1 :: z_0z : z_0z_1$, or

> the ratio of the distance from x_0 to x to the distance from x_0 to x_1
> equals
> the ratio of the distance from z_0 to z to the distance from z_0 to z_1

together with order for points. And proportionality itself can less easily be defined in terms of order for points.

Two special cases are indicated in the adjoining figures.

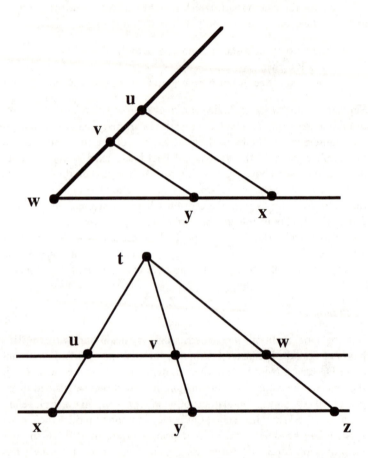

Where *u*, *v*, *w* and *w*, *x*, *y* are collinear (but *u*, *w*, *x* are not), the condition '*wv* : *wu* :: *wy* : *wx*' amounts to the condition that *vy* is parallel to
ux; while where *u*, *v*, *w* are collinear and *x*, *y*, *z* are collinear and *uv* is
parallel to *xy*, the condition '*uw* : *uv* :: *xz* : *xy*' amounts to the condition
that the lines extending *ux*, *vy*, and *wz* meet at a common point. The
general case can be reduced to a combination of these two special cases.
    This purely geometrical definition of proportionality is one of the main
improvements of nineteenth-century over earlier synthetic geometry. The
ancient definition, attributed to Eudoxus, and found in Euclid, presupposes the notions of natural number, counting, and so forth. Or at least,
it makes use of **finite comparative cardinality quantifiers**:

[there are finitely many Fs and finitely many Gs and]
there are as many Fs as there are Gs

The present definition is by contrast purely geometrical, employing no arithmetical or numerical notions.

For another instance, the counterpart of the first coordination primitive:

the horizontal coordinate of x with respect to u, v, w equals
the ratio of the distance from y_0 to y to the distance from y_0 to y_1

can be defined by the condition:

$uz : uv :: y_0 y : y_0 y_1$ where
z is the projection of x in the direction of the line through u, w to
the line through u, v

where the important notion of **projection** here can be defined by:

z is the intersection with the line through u, v of
the unique line parallel to the line through u, w and passing
through x

Second, it can be shown that only natural new axioms are needed, resembling those illustrated in figures. All the new axioms of T^+ are deducible using elementary logic from the axioms of an affine basic pure plane geometry $T^\S$ set forth by Tarski and Szczerba (1965). The distributive law, for instance, roughly speaking corresponds to Desargues's Theorem cited in section 2. This last theory $T^\S$, in the language $L^\S = L^{\dagger\circ}$ having order on points as its only primitive, provides an elegant invariantization and denumericalization of the original T^+ in $L^\dagger$.

4. THE STRATEGY FOR CLASSICAL PHYSICS

a. From Algebra to Analysis

The adaptation of the method of section 3 from the case of a very fundamental level of mathematical geometry can be extended to the case of classical physics in several stages. It can be adapted to the case where one has available in the original, coordinate theory not just the apparatus of algebra but also the further apparatus of analysis. This further apparatus consists of one additional primitive (with attendant additional axioms), for integrity. The method just outlined provides an invariantization and denumericalization with some further apparatus, consisting of one additional primitive (with attendant additional axioms) for what again may be called **integrity**:

$[x_0, x_1, x$ are collinear and x_1 lies between x_0 and x and]
the ratio of the distance from x_0 to x to the distance from x_0 to x_1
   is integral

As this new notion may to some tastes look artificial from a geometric standpoint, it may be desirable to mention alternatives.

Note first that if $x_0x : x_0x_1$ is a strictly positive integer, then the region η consisting of all points y between x_0 and x inclusively such that $x_0y : x_0x_1$ is a non-negative integer contains x_0, x_1, and x and fulfils certain conditions. Namely, it is **discrete** in the sense that:

for each y in η except x there is a y^+ in η
   that is closest to y in the direction of x
for each y in η except x_0 there is a y^+ in η
   that is closest to y in the direction of x_0

And it is **evenly spaced** in the sense that:

for each y in η strictly between x_0 and x,
   the distance from y to y^- equals the distance from y to y^+

And finally, $x_0{}^+$ is just x_1. (It is a fairly easy and pleasant exercise to show that the restricted notion of equidistance used here, for two pairs of points *with all four points involved collinear*, can be defined in terms of parallelism and hence in terms of order.)

Note second that, conversely, if there is a region η fulfilling all the conditions above, then $x_0x : x_0x_1$ is a positive integer. It follows that introduction of the integrity primitive with attendant additional axioms could be avoided in favour of introduction of regions of points with the incidence primitive and appropriate additional axioms (the analogues of extensionality, comprehension, and choice for sets of points). This alternative may to some tastes seem more natural from a geometric standpoint.

The assumption of arbitrary regions, moreover, is not needed. One can make do with any of a number of classes of special regions. One such class is that of finite regions (which are all that are required for the above definition). Another such class is that of open regions (though this would require a slight modification of the above definition), and yet another is that of closed regions. Here a point x is in the **interior** of an interval ab if it is strictly between a and b; an **open** region X in the line is one such that for every point x in X, there are a, b such that x is in the interior of the interval ab and every point in the interior of ab is in X; a **closed** region in the line is one whose complement is open. (Similar definitions can be made for the plane, beginning with the definition that x is in the interior of a triangle abc if there is a point y in the interior of the interval *ab* such that *x* is in the interior of the interval *cy*.) Conversely, it is known that the theories of various special classes of regions are equivalent to each other and to the theory of points with an integrity primitive, in the sense that any one can be reduced to any other by the method of Chapter I.B. Any one of these theories may be taken as the official version of affine intermediate pure plane geometry.

b. From Lower to Higher Dimensions

The method of section 3 can be adapted to any finite number of dimensions. In adapting the method from two-dimensional plane geometry to three-dimensional solid geometry and then four-dimensional hyperspace kinematics, the number of points making up a sequence of benchmarks must be increased from three to four and then five.

c. From Affine to Euclidean

Also at a very basic level of mathematical geometry comes **Euclidean** basic coordinate plane geometry. This is an extension of affine basic coordinate plane geometry with one new primitive for the notion of **equidistance** (for any two pairs of points, not just for the case where all four points involved are collinear):

> *x* lies from *y* as far as *z* lies from *w*, or
> the segment *xy* is congruent to the segment *zw*

and with one new axiom:

> for any points *x*, *y*, *z*, *w* with coordinates
> $X_1, X_2, Y_1, Y_2, Z_1, Z_2, W_1, W_2$ (on the preferred coordinate system),
> *x* lies from *y* as far as *z* lies from *w* if and only if
> $$(Y_1 - X_1)^2 + (Y_2 - X_2)^2 = (W_1 - Z_1)^2 + (W_2 - Z_2)^2$$

The most important of several further notions expressible in terms of equidistance is **perpendicularity**:

> *z* lies from *x* right-angled to how *y* lies from *x*, or
> the segment *xz* is orthogonal to the segment *xy*

expressible by:

> there is a *w* such that *x* lies between *w* and *y* and
> *x* lies from *w* as far as *x* lies from *y* and
> *z* lies from *w* as far as *z* lies from *y*

The method of section 3 can be adapted from the affine to the Euclidean case (and the extension of the method from algebra to analysis, and from lower dimensions to higher dimensions, carries over as well). Indeed, no change in the method at all is needed at the stage of denumericalization. That a change is needed at the stage of invariantization follows from the fact that, there being more axioms, it will be harder for a coordinate system to be admissible in the sense that all the axioms are true on it, so there will be fewer admissible coordinate systems. To put the matter another way, there will be fewer admissible transformations of the preferred coordinate system: in the jargon used in textbooks of linear algebra, the 'Euclidean group' of transformations is smaller than the 'affine group'. Or to put the matter yet another way, it will be harder for a triple of points u, v, w to be admissible in the sense of being the benchmarks of an admissible coordinate system. One still needs non-collinearity. It can be shown that what will be needed in addition for admissibility is the following:

> the segment xz is congruent and orthogonal to the segment xy

With this single change, however, the method of invariantization can be shown to carry over, appropriate axioms being set forth in the semi-popular exposition of Tarski already cited.

d. From Mathematical to Physical

The kinematics of pre-relativistic gravitational theory is intermediate between affine and Euclidean. In the usual jargon, the **'Galilean group'** is intermediate between the affine group and the Euclidean group. To put the matter in another way, more useful in the present context, one does not have a full equidistance notion, but one does have a **partial equidistance** notion:

> x, y, z, w are simultaneous and
> x lies from y as far as z lies from w

with the axiom:

> for any points x, y, z, w with coordinates
> $X_1, X_2, X_3, X_4, Y_1, Y_2, Y_3, Y_4, Z_1, Z_2, Z_3, Z_4, W_1, W_2, W_3, W_4,$
> x, y, z, w are simultaneous and
> 　x lies from y as far as z lies from w if and only if
> $X_4 = Y_4 = Z_4 = W_4$ and

$$(Y_1 - X_1)^2 + (Y_2 - X_2)^2 + (Y_3 - X_3)^2 =$$
$$(W_1 - Z_1)^2 + (W_2 - Z_2)^2 + (W_3 - Z_3)^2$$

The most important of several further notions expressible in terms of equidistance are **simultaneity** itself and **partial perpendicularity**:

> x, y, z are simultaneous and
> z lies from x right-angled to how y lies from x

As in the Euclidean case, the only important change needed from the affine case is in the definition of admissibility of benchmarks u_0, u_1, u_2, u_3, u_4. It can be shown that what will be needed in addition for admissibility is the following:

> u_0, u_1, u_2, u_3 are simultaneous and
> u_0u_1, u_0u_2, u_0u_3 are pairwise congruent and orthogonal and
> u_0, u_4 are non-simultaneous

Otherwise the method of section 3 carries over unchanged. The appropriate axioms can be patched together from the affine and Euclidean cases. They include the assertions that all points constitute under the order relation a four-dimensional affine space, that simultaneity is an equivalence relation, and that all points simultaneous with a given point constitute under the order relation and the partial equidistance relation a three-dimensional Euclidean space.

e. From Kinematics to Dynamics

Classical gravitational theory adds to non-relativistic kinematics certain dynamical notions. Perhaps the most elegant analytic formulation would involve two new notions, **mass density** and **gravitational potential**. These are straightforwardly presentable as, respectively, a scalar field or assignment of real numbers to points, and a vector-field or assignment of triples of real numbers to points (or triple of assignments of real numbers to points). To put the matter another way, one would have, in an analytic formulation, four new primitives as follows:

> the mass density at x is X
> the gravitational potential at x in the ith axial direction is X
>   (for $i = 1, 2, 3$)

An additional arbitrary choice, beyond that of coordinate system, is involved here, namely the choice of scale of measurement for mass. Hence there will be additional complications at the stage of invariantization.

These complications can be unravelled. In the course of invariantizing, the above primitives are replaced by:

the ratio of the mass density at x to the mass density at y is Z

the ratio of the gravitational potential at x
in the direction from y to z
to the gravitational potential at x'
in the direction from y' to z' is W

In the course of denumericalizing, these primitives are replaced by:

the ratio of the mass density at x to the mass density at y
is equal to $z_0 z : z_0 z_1$

the ratio of the gravitational potential at x
in the direction from y to z
to the gravitational potential at x'
in the direction from y' to z' is $w_0 w : w_0 w_1$

Since with most instruments the operational determination of intensive magnitudes like mass reduces to the determination of a ratio of extensive magnitudes of length (the ratio of the distance from the null-mark on the dial to the unit-mark to the distance from the null-mark to the location of the pointer), these should not seem too artificial or devious.

Otherwise the method of section 3 carries over unchanged, and there seems to be no obstacle to extending it to a more comprehensive scientific theory incorporating the classical, Newtonian theory as its fundamental physics. (Of course, in a more comprehensive scientific theory, including natural history in addition to fundamental physics, many particular places and occasions will be mentioned, and there will be no question of invariance for the theory as a whole: the requirement of invariance makes sense only for the fundamental physics.)

f. From Pre-Relativistic to Special-Relativistic

Nor does there seem to be any obstacle to adapting the method further from theories incorporating a non-relativistic perspective to theories incorporating a special-relativistic perspective, or in the usual jargon, involving the '**Lorentz–Poincaré group**' rather than the Galilean group. To be sure, in the former case, a pure or 'synthetic' geometry appropriate to the non-relativistic case is obtainable by patching together pure or 'synthetic' geometries appropriate to the affine and Euclidean cases, and these were made available in the work beginning with Euclid's, and hence long before the coordinate or 'analytic' geometry was made available by the work of Descartes and Pierre de Fermat. Whereas in the latter case, the coordinate or 'analytic' geometry appropriate to the special-relativistic case came first, in the work of Hermann Minkowski, so that one speaks of **Minkowski** space. Nonetheless, a pure or 'synthetic' geometry appropriate to the special-relativistic case, comparable in style to Euclid's (as rigorized by Hilbert) while agreeing in substance with Minkowski's, was made available shortly afterwards in the work of Alfred Robb, see Robb (1914). A more formalized version of this little-known work of Robb could play the role in the special-relativistic case that the better-known work of Tarski and his school has played in the non-relativistic case.

5. OBSTACLES TO EXTENDING THE STRATEGY

a. Post-Classical Physics

Obstacles that it is not known how to surmount arise in connection with quantum mechanics. The difficulty is not with the underlying geometry or kinematics, which is Minkowskian or special-relativistic. Rather, the difficulty is with the additional physical or dynamical notions involved. The difficulty is that when quantum-mechanical theories are presented in field-theoretic form—and it is presumably this form, rather than the particle-theoretic form, that one would want to consider—the mathematical objects that have to be assigned to points are much more complicated than scalars (single real numbers) or three- or four-vectors (triples or quadruples of real numbers). It is known (and was indicated in article I.B.1.b) how to represent these more complicated mathematical objects by real numbers. But the known representation is devious and awkward, not straightforward and elegant.

Further obstacles arise in connection with general relativity, and these do arise from the underlying geometry, which is Riemannian or differential. At the stage of invariantization, there are difficulties including the following. In all of the cases considered previously, it was possible to avoid quantifying over coordinate systems, and so ascending to a higher level of abstraction, because coordinate systems were representable by their benchmark points. But this is not so in the more general coordinate systems of Riemannian or differential geometry, and the usual approach to invariantizing in this case does involve ascending to a higher level of abstraction, which would seem to create major obstacles to subsequent denumericalization. Even if these difficulties were surmounted, at the stage of denumericalization there would be the further and major difficulty that there seems to be no pure or 'synthetic' Riemannian or differential geometry, no list of axioms that are natural-looking from a geometric viewpoint and that could play the role in this case that was played by the lists of axioms provided by the work of Tarski and his school in most of the cases considered previously (or of Robb in the special-relativistic case).

As for a physical theory unifying quantum mechanics and general relativity, such a theory could be expected to present both the difficulties presented by the former, and the difficulties presented by the latter, and also new difficulties of its own. But the most obvious obstacle to developing an elegant, synthetic, pure, natural-looking, invariant, straightforward version of such a theory at present is the circumstance that so far no one has developed even an inelegant, analytic, coordinate, artificial-looking, arbitrary-choice-dependent, devious version of such a theory. Thus a question mark hangs over the geometric nominalist strategy. But two more positive remarks are in order.

First, if one does care about realism, but does not care about elegance, one can probably get by just routinely applying the general method of Chapter I.B. Essentially all that is required is that there be some kind of 'metric' $\mu(x, y)$ or measure of separation between points x, y, and that the geometry assume that space is continuous rather than discrete. For this amounts to assuming that every real number can be represented as a ratio $\mu(x, y) : \mu(u, v)$, which is enough to make the general method applicable. Second, to say that it is not known to be possible to provide elegant nominalistic versions of quantum mechanics and general relativity is not to say that it is known to be impossible. Rather, whether the obstacles enumerated can be surmounted is an open research problem. As a consequence of nominalism's being mainly a philosopher's concern, this open research problem is moreover one that has so far been investigated only by amateurs—philosophers and logicians—not professionals—geometers and physicists; and the failure of amateurs to surmount the obstacles is no strong grounds for pessimism about what could be achieved by professionals.

b. Non-Empirical Physics

According to substantivalism, geometric entities are to be recognized as physical entities and geometric relationships as physical relationships, and hence presumably geometric questions as physical questions. Now, when it was claimed in article I.B.1.b that analysis without any higher set theory provides enough mathematics for applications in physics, 'physics' was being understood in the ordinary sense rather than this substantivalist sense. So it must now be asked, first whether analysis without any higher set theory provides enough mathematics for applications to geometry; and second whether, if not, that fact poses a problem for the nominalistic strategy of this chapter. The first question is technical, the second philosophical.

To state the technical question more precisely, consider the kind of geometric theory T used in the strategy of this chapter, intermediate affine pure geometry as in article 3.a. (As mentioned there, it may be formulated either in terms of points and a special 'integrity' primitive, or in terms of points and certain special classes of regions. The latter formulation will be more appropriate to present purposes.) Let T^+ be the result of adding the apparatus of analysis to T, and $T^!$ the result of adding the apparatus of set theory. The question whether analysis is sufficient for applications to geometry and hence for 'physics' in the substantivalist sense, or whether higher set theory is needed, amounts to the following question: is there any geometric assertion P, any assertion expressible in the language L of T, whether or not it would have obvious implications for 'physics' in the ordinary sense, that can be proved in $T^!$ but that cannot be proved in T^+? It has already been shown in this chapter that any assertion P expressible in L and provable in T^+ is in fact provable in T, since the apparatus of analysis can be reconstrued geometrically. Hence the question may be reworded: is there any conjecture P expressible in the language L of T but not provable in T, that becomes provable in $T^!$?

The answer is yes: by the **Incompleteness Theorem** of Kurt Gödel (together with what has been shown in this chapter, that the apparatus of analysis can be reconstrued geometrically), it follows that the assertion that the theory T is consistent can be reconstrued or coded geometrically as a conjecture $P(T)$ expressible in the language L of T, and that it is not provable from the axioms of T. It is, however, provable from the axioms of $T^!$. However, three remarks are in order. First, as a coded assertion, $P(T)$ is not at all natural as an assertion of geometry. Further, as a consistency assertion, $P(T)$ is something that no one who genuinely accepts the theory T genuinely could doubt. Finally, the Gödel method applies to any theory, so for instance the assertion that the theory $T^!$ is consistent can also be reconstrued or coded geometrically as a conjecture $P(T^!)$ in the language L of T and it cannot be proved in $T^!$. These remarks might be summed up by saying that the Gödel example is one where adding set theory to geometry makes only a 'formal' difference. There are, however, other examples where it makes a 'material' difference.

To recall the philosophical question, however, would even an example that makes a 'material' difference pose a problem for the nominalistic strategy of this chapter? This question may be reformulated: suppose there is some natural geometric question P that cannot be settled by proof or disproof in the nominalistically acceptable theory T, but can be settled in the nominalistically unacceptable theory $T^{\sharp}$ that results when the apparatus of set theory is added; is that a problem for the nominalist? The answer is, not unless and until the geometric theorem comes to play a role in physics in the ordinary sense of 'physics' and not just in the substantivalist sense of 'physics'. For only if and when that happens is the nominalist obliged to concede that the question that has been settled by set theory has been settled the right way. And only if and when the nominalist is obliged to concede that set theory settles geometric questions that cannot be settled nominalistically and settles them the right way, does the nominalist's position come to resemble the instrumentalism that it is the aim of strategies like that of this chapter to avoid. And so far, examples where set theory makes a 'material' difference to geometry have not yet come to play a role in 'physics' in the ordinary sense. Still, it may be of interest to describe briefly one such example.

The first task in presenting examples is to introduce the special classes of regions, and the special properties of regions, involved. The definitions will be given for the line; the corresponding definitions for the plane are exactly analogous. The special classes of **open** regions, and their complements the **closed** regions, have already been defined. The next important special classes of regions are the F_σ regions, and their complements the G_δ regions. The usual analytic definition is that an F_σ region is one that is a union $F_0 \cup F_1 \cup F_2 \cup \ldots$ of countably many closed regions. It is important in the present context to note that a purely synthetic definition is available, using only geometric notions there has already been reason to introduce, such as that of evenly spaced points in a line or of projection of a point in the plane to the line. Derivatively, parallel lines in the plane are evenly spaced if their points of intersection with any transverse line are evenly spaced, and the projection of a region in the plane to the line is the linear region whose points are the projections of the points of the planar region. A region in a line X is F_σ if it is the projection to X of the intersection of two regions $Y \cap Z$, where Y is a closed region in the plane, and Z is a region in the plane that is a union of evenly spaced lines parallel to X. The last important special classes of regions are the

analytic regions, the projections to the line of G_δ regions in the plane; the **co-analytic** regions, their complements; and the **Borel** regions, the regions that are both analytic and co-analytic. (The importance of the analogues of Borel regions of linear points, namely, Borel sets of real numbers, in the mathematical apparatus of sophisticated physics was mentioned in article I.B.1.b.)

The second task in presenting examples is to present the special properties of regions involved. The pertinent properties are the ones usually explained intuitively as follows. Consider a horizontal interval I in the line. For any subregion A of I one may imagine an infinite game for two players, IN and OUT. I consists of two halves, left and right. The game begins with IN picking one of these. It in turn consists of two halves. The game continues with OUT picking one of these. It in turn consists of two halves. The game continues with IN picking one of these. Alternating in this way, IN and OUT successively pick smaller and smaller intervals, which in the infinite limit narrow down to a single point. IN or OUT wins according as this point is in or out of A. A **strategy** (respectively, **counter-strategy**) is a rule telling IN (respectively, OUT) what to pick at each stage, as a function of the opponent's previous picks. A strategy (respectively, counter-strategy) is **winning** if when IN (respectively, OUT) picks according to the rule it provides, IN (respectively, OUT) always wins. The game is called **determined** and the set A is called **determinate** if there is either a winning strategy or a winning counter-strategy. It is important in the present context to note that a purely synthetic definition is available. Consider any horizontal interval J. It may be bisected into two equal subintervals J', J''. There is a rectangle R' with upper horizontal side J' and with vertical side half as long as its horizontal side, and a similar rectangle R'' for J''. Call these the dependent rectangles of J. Consider now the square whose lower horizontal side is I. The primary and secondary subrectangles of this square are defined as follows. The rectangles dependent on the upper horizontal side of the square are primary, and called the initial rectangles. For any primary rectangle R, the dependent rectangles on its lower side are secondary rectangles, called the successor rectangles of R. For any secondary rectangle R, the dependent rectangles on its lower side are primary rectangles, called the successor rectangles of R. A strategic region is one that contains exactly one initial rectangle and that, whenever it contains a primary rectangle R, contains each successor of R and exactly one successor of each successor of R. A counter-strategic region is analogously defined. Parts of such a set are illustrated in the adjoining figure, which invite comparison with the

simpler among the computer graphics of 'fractals' produced by Benoît Mandelbrot, in which a similar kind of structure is endlessly reproduced on smaller and smaller scales.

The intersection of a strategic region S' and a counter-strategic region S'' consists of a single large rectangle at the top, a single smaller one next below it, and a single smaller one next below that, and so on, and contains only one complete vertical interval of unit length. Its lower boundary point is a point in the interval I called the outcome of S', S''. A subregion A of I is positively (respectively, negatively) determinate if there is a strategic S' (respectively, a counter-strategic S'') such that for any counter-strategic S'' (respectively, any strategic S'), the outcome of S', S'' is in (respectively, out of) X. X is determinate if it is either positively or negatively determinate.

The hypotheses **GD**, **BD**, and **CD** of open, **Borel**, and co-analytic determinacy are respectively that every open, every Borel, or every co-analytic region is determinate. The status of these geometric hypotheses in relation to set theory is as follows. Specialists in set theory have considered adding further axioms to the axioms ZFC accepted by mathematicians generally. Roughly speaking, most of those who favour adding new axioms favour something called the axiom of **measurables**, which produces a system ZFM; a few may favour something called the axiom of **constructibility**, which produces a system ZFL. (The two axioms, of measurables and of constructibility, are incompatible.) So besides the system $T^\ddagger$ obtained by adding the apparatus of ZFC to the geometric theory T, one can consider the systems $T^\#$ and $T^\pm$, both stronger than $T^\ddagger$, and incompatible with each other, obtained by adding ZFM and ZFL. Then GD can be proved in the geometric theory T (Gale and Stewart). BD can be proved in $T^\ddagger$ (D. A. Martin), but not in T (Harvey Friedman). CD can be proved in $T^\#$, but not in $T^\ddagger$, since it can actually be disproved in $T^\pm$ (Martin, Friedman again). Some further information is available in Burgess (1989). For a philosophically oriented account of these and related matters, the reader is referred to the work of Penelope Maddy, especially the two-part paper Maddy (1988), which contains full references to the original technical literature, and which makes clear, as the present brief account cannot, the central role of determinacy in the qualitative (as opposed to quantitative) theory of regions of linear points, or in the more usual analytical terminology, the descriptive (as opposed to metric) theory of sets of real numbers, or descriptive set theory for short.

B

A Purely Modal Strategy

0. OVERVIEW

The label 'nominalism' by its very etymology suggests a view that would avoid abstracta in favour of their names, for instance avoiding numbers in favour of numerals. An obstacle has already been indicated (in article I.B.2.a), namely, the fact that there aren't enough concrete tokens of numerals; and a solution has already been hinted at (again in article I.B.2.a), to consider not just what numerals there are, but also what numerals there could have been. This is the strategy of the simplest version of modal nominalism. It is purely modal in that no other extended logics beyond modal logic are employed.

This purely modal nominalism, like geometric nominalism, has two tasks, the technical one of implementing the strategy, and the philosophical one of defending its apparatus. That apparatus faces criticism from two sides. On one side, the modal nominalist must defend **intensionalism**, or acceptance of modal logical notions, against **extensionalism**, according to which modal notions should be avoided as obscure and confused. On the other side, modal nominalism must defend **primitivism**, acceptance of modal logical distinctions as undefined, against **reductivism**, which would reconstrue modal logical notions in terms of an apparatus of unactualized possibilia.

Reductivism would reconstrue the assertion to the effect that there possibly could have been some things there actually aren't into the assertion that there are some unactualized possible things in some unactualized possible world. Now one can quibble over whether or not the unactualized possible inhabitants of an unactualized possible world should be called 'abstract', since they may be causally active in the sense that they causally interact *with each other*; but since they are causally isolated in the sense that none of them causally interact *with us*, such exotica as unactualized possibilia are quite as repugnant to nominalists as are numbers or sets. That is why a modal nominalist must be a primitivist.

There are further subdivisions among the two groups of critics: the extensionalists are divided into moderates and extremists, and there is a distinction to be made between **revolutionary** and **hermeneutic** reductivists. The extreme extensionalist claim that modal locutions 'are simply unintelligible' or 'just cannot be understood' will be dismissed here with the observation that that claim itself involves the very modal locutions ('-ible', 'cannot') it pretends to reject. This observation is merely a specific instance of the more general one that modal locutions are ubiquitous in everyday discourse and in the 'soft' sciences. Historically, modal locutions have been much used even in the 'hardest' science of all, mathematics: as Paul Bernays remarked, at the beginning of Bernays (1935), contrasting the axiomatics of Euclid with that of Hilbert, where the modern practice is to speak in axioms of what 'there exists', the ancient practice was to speak of what 'it is possible to produce'. Modal nominalism is thus, like geometric nominalism, a proposal to turn back the clock. Historical relics of the former practice persist even today in colloquial mathematical language, as when one speaks of an integer or equation or function or space or formula that is divisible or solvable or differentiable or metrizable or provable, as being one that can be divided or solved or differentiated or metrized or proved—whereas the formal definition is that it is one for which there exists a divisor or solution or derivative or metric or proof.

The moderate extensionalist claim is that in mathematics and mathematically formulated 'hard' science, modality has been eliminated in favour of abstracta, and that this replacement has constituted significant clarification and significant progress. The revolutionary reductivist claim is similarly that the general replacement of a primitive modal sentential operator ('it could have been the case that . . .') by a novel apparatus of existential quantification over possible worlds ('in some possible world it is the case that . . .') would constitute significant clarification and significant progress. The modal nominalist counter-claim is that in both cases one has not progress but its opposite, on account of the abstracta or exotica introduced. The grounds for this counter-claim depend on just what the grounds for sympathy with nominalism are supposed to be in the first place.

The hermeneutic reductivist claim is roughly that the possible worlds are not novel apparatus, but rather that involvement with them is already present, though latent rather than manifest, in ordinary talk of what could or would or might have been the case. A somewhat fuller formulation of the hermeneutic reductivist claim will be given in section 1 below.

(Also, to some extent the later discussion of hermeneutic nominalism, mentioned in section I.A.o and deferred until the Conclusion of this book, may apply *mutatis mutandis* to hermeneutic reductivism.)

Even when general criticisms are dismissed, it remains important to clarify the particular apparatus deployed in any particular strategy. This is especially so because there is no one universally accepted system of modal logic: the standard literature of modal logic contains scores of competing options. Moreover, the modal logic appropriate to the particular strategy of this chapter is not even one of the usual candidates. In its technical task, whereas geometric nominalism found much that was useful for its purposes in the classical literature on the geometric algebra, modal nominalism finds little of use in the existing literature on modal logic, and has to proceed largely independently.

Nothing helps better to clarify modality than the analogy with temporality. This analogy will be used throughout the informal discussion of modality in general in section 1, and of the modal apparatus deployed in the strategy of this chapter in particular in section 2. A more formal discussion of the modal logic appropriate to the strategy will be postponed until article 3.b, after presenting the essentials of the strategy in article 3.a. (Technicalities are relegated to the optional semi-technical appendix, article 3.c.)

1. MODAL NOTIONS: GENERALITIES

a. Temporal

The simplest distinctions of temporality are expressed in natural languages like English through tense (past, present, future), a system of verbal inflections and auxiliaries. Slightly more complicated distinctions are expressed using nouns, which bring with them perhaps an appearance of existential assumptions: 'at that time' or 'at some time' or 'at the same time'.

b. Modal

The simplest distinctions of modality are expressed in natural languages like English through mood (indicative, subjunctive, conditional), a system of verbal inflections and auxiliaries. Slightly more complicated distinctions are expressed using nouns, which bring with them perhaps an appearance of existential assumptions: 'in that contingency' or 'in some contingency' or 'in the same contingency'.

Scientific thought makes use of a systematic analogy between (and finally, in relativity, a merger of) the temporal and the spatial. For instance, duration is measured analogously to length. In analogy to the notion of a maximally specific place, or point, the notion of a maximally specific time, or **instant**, is introduced, and perhaps along with it the notion of a temporary **stage** of the world at some instant —or **temporary world** in a more extravagant usage.

The fullest development of this analogy brings with it a new tenseless way of speaking. Sentences, true at some times and not at others:

(i)   The body is charged

are replaced by one-place predicates of such **index** entities as times or instants or stages:

(ii)   The body [is] charged at instant *t*

The brackets here are supposed to indicate that the verb is really tenseless, not present-tense. Likewise *k*-place predicates are replaced by (*k* + 1)-place predicates, with an extra place for index entities. A sentence involving tense:

(iii)   There was a body that was charged

Scientific thought makes use of a systematic analogy between the modal and the spatial. For instance, probability is measured analogously to length, area, or volume. In analogy to the notion of a maximally specific place, or point, the notion of a maximally specific contingency, or **case**, is introduced, and perhaps along with it the notion of a possible **state** of the world in some case—or **possible world** in philosophical usage.

The fullest development of this analogy brings with it a new moodless way of speaking. Sentences, true in some contingencies and not at others:

(i)   The particle spins downwards

are replaced by one-place predicates of such **index** entities as contingencies or cases or states:

(ii)   The particle {spins} downwards in case *u*

The braces here are supposed to indicate that the verb is really moodless, not indicative-mood. Likewise *k*-place predicates are replaced by (*k* + 1)-place predicates, with an extra place for index entities. A sentence involving mood:

(iii)   There could have been a particle that spun downwards is replaced by a tenseless one:

(iv)　　There [is] an instant, earlier than the present, at which there [is] a body and it [is] charged

Talk of the changes in a body's condition at various times is replaced by talk of timeless relations between the body and different index entities.

Temporality is not present in mathematics. (Recall the discussion in article I.A.1.b: one does not ask of a prime number, 'When did it become prime?') The mathematical modelling of physical phenomena thus inevitably involves the kind of de-tensing just described. In its simplest form, the method of treating time as an extra dimension or coordinate has by now become familiar even to most lay people from the representation of the motion of a particle by the graph of its trajectory. These methods have by now reached quite complex forms in the mathematics of general relativity and dynamical systems.

Modern logic was initially developed for purposes of analysing mathematical arguments. Hence no provision was made in it for tense. If one wishes to apply it to arguments turning on tense, the simplest procedure is just to regiment by de-tensing as just described. Thus it is that arguments involving steps like (iii) are regimented, symbolized, and formalized, result-

is replaced by a moodless one:

(iv)　　There {is} a case in which there {is} a particle and it {spins} downwards

Talk of the chances of a particle's behaviour in various contingencies is replaced by talk of moodless relations between the body and different index entities.

Modality is not present in modern mathematics. (Recall the discussion in article I.A.1.b: one does not ask of a prime number, 'What if it hadn't been prime?') The mathematical modelling of statistical phenomena thus inevitably involves the kind of de-mooding just described. In its simplest form, the method of spatialization of a range of contingencies or cases has become familiar even to many lay people from Euler or Venn diagrams. These methods have by now reached quite complex forms in the mathematics of statistical thermodynamics and quantum mechanics.

Modern logic was initially developed for purposes of analysing mathematical arguments. Hence no provision was made in it for mood. If one wishes to apply it to arguments turning on mood, the simplest procedure is just to regiment by de-mooding as just described. Thus it is that arguments involving steps like (iii) are regimented, symbolized, and formalized, resulting in arguments involving steps like:

(v) $\exists t \exists x (t < t_\circ \wedge Ctx)$

Here the $\exists$ is tenseless, and the variable t ranges over instants, and the variable x over bodies that may not exist at present.

Such regimentation has seemed artful in a complimentary sense to some, including Quine, and artificial in a derogatory sense to others, notably Arthur Prior, who from largely though not wholly linguistic motives pioneered the development of an **autonomous temporal** or tense logic. This logic enriches elementary logic with operators $\mathcal{P}$ and $\mathcal{F}$, for past and future.

> 'it has been the case that . . .'
> 'it will be the case that . . .'.

In this logic, instead of (iv), (v) one would have:

(vi)   it was the case that (there is a body x such that (x is charged))

(vii) $\mathcal{P}\exists x\ Cx$

This formalism, with just the operators $\mathcal{P}$ and $\mathcal{F}$, is not as expressive as that allowing explicit quantification over indices. That is not necessarily a disadvantage, if the formalism is expressive enough, and if it is more manageable than a more expressive formalism would be. Tense logic and related logics have

ing in arguments involving steps like:

(v) $\exists u \exists y Duy$

Here the $\exists$ is moodless, and the variable u ranges over cases, and the variable y over bodies that may not actually exist.

Such regimentation has seemed artful in a complimentary sense to some, including D. K. Lewis, and artificial in a derogatory sense to others. The development of an **autonomous model** or mood logic, which provides an alternative, was pioneered by C. I. Lewis. This logic enriches elementary logic with operators $\Diamond$ and $\square$, for possibility and necessity:

> 'it could have been the case that . . .'
> 'it couldn't have not been the case that . . .'

In this logic, instead of (iv), (v) one would have:

(vi)   possibly (there is a particle y such that (y spins downward))

(vii) $\Diamond \exists y Dy$

This formalism, with just the operators $\Diamond$ and $\square$, is not as expressive as that allowing explicit quantification over indices. That is not necessarily a disadvantage, if the formalism is expressive enough, and if it is more manageable than a more expressive formalism would be. Modal logic and related logics been found useful for some studies in theoretical computer science for just this reason. Additional operators can be introduced. (For examples see article 3.b.) And if one does not care about whether the operators have natural readings in English, the full expressive power of quantification over indices can be recovered by introducing enough operators. (In this connection see article III.B.3.c.)

The formalism is also not very faithful to English grammar, which allows only limited iteration of tenses, whereas tense logic allows unlimited iteration of P and F. However, autonomous temporal logics do banish such exotica as index entities and eventual-but-not-present, sooner-or-later-but-not-now entities from their 'syntax' or proof theory.

Where both are applicable, canonical formalization in terms of standard logic and instants, and autonomous formalization in an extended logic with primitive temporal sentential operators, assign very different logical formalizations to the premisses and to the conclusion of an argument, the autonomous formalization being closer to the surface grammatical form. However, the two schemes of formalization agree completely in their judgements as to whether the premisses imply the conclusion.

Along with a proof theory providing methods of deduction for showing that conclusions do follow

have been found useful for some studies in theoretical computer science for just this reason. Additional operators can be introduced. (For examples see article 3.b.) If one does not care about whether the operators have natural readings in English, the full expressive power of quantification over indices can be recovered by introducing enough operators. (In this connection see article III.B.3.c.)

The formalism is also not very faithful to English grammar, which has no compound moods, whereas modal logic allows unlimited iteration of $\Diamond$ and $\Box$. However, autonomous modal logics do banish such exotica as index entities and possible-but-not-actual, might-have-been-but-aren't entities from their 'syntax' or proof theory.

Where both are applicable, canonical formalization in terms of standard logic and worlds, and autonomous formalization in an extended logic with primitive modal sentential operators, assign very different logical formalizations to the premisses and to the conclusion of an argument, the autonomous formalization being closer to the surface grammatical form. However, the two schemes of formalization agree completely in their judgements as to whether the premisses imply the conclusion.

Along with a proof theory providing methods of deduction for showing that conclusions do follow from premisses, there is wanted a model theory providing methods of counter-example for showing that conclusions do not follow. (A counter-example would be a model in which the premisses are true but not the conclusion.) And here a problem arises.

Ideally, one would hope for completeness, for a model theory in which there is always a counter-example if the conclusion is indeed not deducible from the premisses. (This is achieved for the usual model theory for standard logic by the Completeness Theorem, as cited in article I.B.4.a. It cannot be achieved for some extended logics, such as those considered in the next chapter.) But one cannot simply imitate the procedure for standard logic as found in standard textbooks.

One cannot simply adapt the usual clause:

> (~*Q*) is true in *M* if and only if *Q* is not true in *M*

and write:

> (*PQ*) is true in *M* if and only if *Q* has been true in *M*

For if *M* is a *mathematical* model, nothing has held in *M* that doesn't now hold.

The solution offered by 'Kripke models', which have by now almost completely superseded earlier 'algebraic models', is in effect just to take as models for temporal-logic

from premisses, there is wanted a model theory providing methods of counter-example for showing that conclusions do not follow. (A counter-example would be a model in which the premisses are true but not the conclusion.) And here a problem arises.

Ideally, one would hope for completeness, for a model theory in which there is always a counter-example if the conclusion is indeed not deducible from the premisses. One cannot simply imitate the procedure for standard logic as found in standard textbooks.

One cannot simply adapt the usual clause:

> (~*Q*) holds in *M* if and only if *Q* does not hold in *M*

and write:

> ($\Diamond Q$) is true in *M* if and only if *Q* could have been true in *M*

For if *M* is a *mathematical* model, nothing could have held in *M* that doesn't actually hold.

The solution offered by 'Kripke models', which have by now almost completely superseded earlier 'algebraic models', is in effect just to take as models for modal-logic formulas like (vii) the models of their standard logic counterparts like (v). Thus something like instants appear in the 'semantics' of temporal logic in the logicians' sense of 'semantics' as a theory of models. The claim of hermeneutic reductivism is that they also belong in the 'semantics' of temporal notions in the linguists' sense of 'semantics' as a theory of meaning. (It would, needless to say, be a straightforward fallacy of ambiguity to infer their presence in 'semantics' in the latter sense from their presence in 'semantics' in the former sense.)

formulas like (vii) the models of their standard logic counterparts like (v). Thus something like worlds appear in the 'semantics' of modal logic in the logicians' sense of 'semantics' as a theory of models. The claim of hermeneutic reductivism is that they also belong in the 'semantics' of modal notions in the linguists' sense of 'semantics' as a theory of meaning. (It would, of course, be a simple fallacy of equivocation to infer their presence in 'semantics' in the latter sense from their presence in 'semantics' in the former sense.)

2. MODAL NOTIONS: SPECIFICS

The discussion of this section refers to the following scenario, which illustrates in miniature the kind of modal notions to be deployed in the strategy outlined in section 3. Suppose that today there is a stick of chalk x weighing .02 (in kilograms, and to the nearest gram), and another stick of chalk y weighing .01. Consider the following programme for tomorrow.

11.00 a.m.:
Most of x is used up inscribing a large token X of the numeral '.02' on a blackboard.

11.15 a.m.:
The decimal point in X is erased, and some of y is used up inscribing a new one nearby, leaving a large token X' of the numeral '0.2' on the blackboard.

11.30 a.m.:
X' is erased, and the rest of x and y are used up inscribing a large token Y of the numeral '.01' on the blackboard.

11.45 a.m.:
The blackboard is erased.

a. Temporal

Suppose the above scenario will be acted out tomorrow. What then will (by noon tomorrow) have been the case? The claim seems intuitively plausible that there will have existed token numerals X, X', Y and that:

(i) X will have marked how much x weighs.

(ii) Y will have marked how much y weighs.

(iii) X will have marked more than Y (and less than X') will have marked.

(iv) X' will have been distinct from X.

And the following additional claims seem equally intuitively plausible.

The claim 'there will have existed token numerals . . .' is a claim about the eventual but not present existence of an ordinary sort of entity, not a claim about the existence of an extraordinary eventual-but-not-present sort of entity. (i)–(iv) hold even though none of X, X', Y (now) exists: the occurrences of x, y are in the scope of 'there is a stick of chalk . . .', while the occurrences of X, X', Y are in the scope of 'there will have been a token numeral . . .'. The occurrences of x, y in (i)–(iv), with present tense verbs do, while the occurrences of X, X', Y with future perfect tense verbs don't, involve **existential import**. (i) holds even though X

b. Modal

Suppose that the above scenario will not after all be acted out tomorrow. Still, it could have been. If it had been, what then would have been the case? The claim seems intuitively plausible that there would have existed token numerals X, X', Y and that:

(i) X would have marked how much x weighs.

(ii) Y would have marked how much y weighs.

(iii) X would have marked more than Y (and less than X') would have marked.

(iv) X' would have been distinct from X (and from Y).

And the following additional claims seem equally intuitively plausible.

The claim 'there could have existed token numerals . . .' is a claim about the possible but not actual existence of an ordinary sort of entity, not a claim about the existence of an extraordinary possible-but-not-actual sort of entity. (i)–(iv) hold even though none of X, X', Y (actually) exists: the occurrences of x, y are in the scope of 'there is a stick of chalk . . .', while the occurrences of X, X', Y are in the scope of 'there could have been a token numeral . . .'. The occurrences of x, y in (i)–(iv), with indicative mood verbs do, while the occurrences of X, X', Y with conditional mood verbs don't, involve **existential import**. If solemn jargon is will not have marked how much x will have weighed when X will have existed, since most of x will have been used up in bringing X into existence. (i) involves a **present/ eventual cross-comparison**, a comparison or relation between how something there is is and how something there will have been will have been.

(ii) similarly holds even though Y will not have marked how much y will have weighed when Y will have existed, and indeed y will not have existed when Y will have existed, since y will have been wholly used up in bringing Y into existence. More generally, (i)–(iv) involve no **coexistential import**, no implication or presupposition that X, X', Y will ever have coexisted, no implication or presupposition of **contemporaneousness**.

(iii) again holds even though X and Y will never have coexisted, since X will have been erased before Y will have been brought into existence. (iii) involves an **eventual/eventual cross-comparison**, a comparison or relation between how one thing there will have been will have been and

wanted for the kind of commitment they do involve, commitment as to what is possible or could have existed, this may be called **dynat-ontological commitment**. (i) holds even though X would not have marked how much x would have weighed if and when X would have existed, since most of x would have been used up in bringing X into existence. (i) involves an **actual/hypothetical cross-comparison**, a comparison or relation between how something there is is and how something there would or might have been would or might have been.

(ii) similarly holds even though Y would not have marked how much y would have weighed if and when Y would have existed, and indeed y would not have existed if and when Y would have existed, since y would have been used up in bringing Y into existence. More generally, (i)–(iv) involve no **coexistential import**, no implication or presupposition that X, X', Y would or could ever have coexisted, no implication or presupposition of **compossibility**.

(iii) again holds even though X and Y would never have coexisted, since X would have been erased before Y would have been inscribed. (iii) involves a **hypothetical/hypothetical cross-comparison**, a comparison or relation between how one thing there could have been would or might have been and how another thing there will have been will have been; and this cross-comparison involves no implication or presupposition of contemporaneousness.

how another thing there could have been would or might have been; but this cross-comparison involves no implication or presupposition of compossibility. If solemn jargon is wanted, commitment as to what are compossible or could have coexisted may be called **syndynatontological commitment**. (To be sure, in the example as given X and Y would have coexisted successively, though not contemporaneously, but the example can be modified so that what will be compared are two different token numerals each of which (but not both of which) could have been made from the same chalk differently used.)

(iv) holds even though how a numeral functions linguistically depends on how it is constituted physically, and even though X' will have been only slightly differently constituted physically from X. For numerals that are constituted physically only slightly differently may nonetheless function linguistically quite differently, and so be distinct entities. How a numeral functions linguistically is a **permanent**, not a **temporary** feature of its identity: a numeral either will always have functioned thus-and-so, or will never have done so; it is not the case that one and the same numeral may first function thus-and-so and then not. If an earlier numeral will have functioned linguistically differently from a later numeral, then the latter will have been distinct

(iv) holds even though how a numeral functions linguistically depends on how it is constituted physically, and even though X' would have been only slightly differently constituted physically from X. For numerals that are constituted physically only slightly differently may nonetheless function linguistically quite differently, and so be distinct entities. How a numeral functions linguistically is an **essential**, not an **accidental** feature of its identity: a numeral either would have functioned thus-and-so, or would not; it is not the case that one and the same numeral might have functioned thus-and-so but also might not. A numeral that functioned thus-and-so and one that did not *ipso facto* would not have been one and the same from, and not identical to, the former. In (i), (ii) the present/eventual cross-comparisons involved are thus **present/permanent** rather than **present/temporary**; and in (iii) the eventual/eventual cross-comparison is a **permanent/permanent** cross-comparison.

Certain optional phrases serving as reminders of these glosses may be inserted into (i)–(iii):

(i′) X (while it will have existed) will (always) have marked how much x (now) weighs.

(ii′) Y (while it will have existed) will (always) have marked how much y (now) weighs.

(iii′) X (while it will have existed) will (always) have marked more than Y (while it will have existed) will (always) have marked.

In (i′)–(iii′), the insertion of 'always' (rather than 'once') emphasizes or underscores the permanence of linguistic functioning. The insertion of 'while it will have existed' emphasizes the absence of existential import. In (i′), (ii′) the insertion of 'now' emphasizes that it is a present/eventual cross-comparison that is being made. In (iii′) the separate insertion twice of 'while it will have existed' emphasizes the absence of coexistential import, and that it is an eventual/eventual cross-comparison that is being made. The

numeral. In (i), (ii) the actual/hypothetical cross-comparisons involved are thus **actual/essential** rather than **actual/accidental** cross-comparisons; and in (iii) the hypothetical/hypothetical cross-comparison is an **essential/essential** cross-comparison.

Certain optional phrases serving as reminders of these glosses or scholia may be inserted into (i)–(iii):

(i′) X (if it had existed) would (necessarily) have marked how much x (actually) weighs.

(ii′) Y (if it had existed) would (necessarily) have marked how much y (actually) weighs.

(iii′) X (if it had existed) would (necessarily) have marked more than Y (if it had existed) would (necessarily) have marked.

In (i′)–(iii′), the insertion of 'necessarily' (rather than 'possibly') emphasizes or underscores the essentiality of linguistic functioning. The insertion of 'if it had existed' emphasizes the absence of existential import. In (i′)–(ii′) the insertion of 'actually' emphasizes that it is an actual/hypothetical cross-comparison that is being made. In (iii′) the separate insertion twice of 'if it had existed' emphasizes the absence of coexistential import, and that it is a hypothetical/hypothetical cross-comparison that is being made. The enclosure

enclosure of these insertions in parentheses underscores that they are optional, already implicit or tacit in (i)–(iii), mostly in the sequence of tenses used.

of these insertions in parentheses underscores that they are optional, already implicit or tacit in (i)–(iii), mostly in the sequence of moods used. In particular, that 'necessarily' rather than 'possibly', and hence actual/essential and essential/essential comparison, is intended, can be understood from the use of the modal auxiliary 'would' rather than of 'might'.

3. THE STRATEGY

a. Outline

The starting-point for the strategy to be outlined in this section will be an analytically formulated theory T in a language L (as in article I.B.3.d). The idea will be to replace assumptions about the actual existence of numbers by assumptions about the possible existence of numerals. In section 2 finite numerals for rational numbers were considered, but the strategy will consider infinite numerals for real numbers. Unlike the numerals in section 2 these need not be numerals of the decimal system; they need not be made of chalk; and they need not have been constructed for the purpose of serving as numerals. Numerals may be taken to be any sort of physical entities provided they are sufficiently complex that each can somehow be construed as coding in some physical way an infinite binary or zero-one sequence. The first step in the strategy involves no change in the formal theory, but merely in the intuitive understanding of it. The variables $X, Y, \ldots$ are understood to range not over real numbers, but rather over numerals for them. The primitives are understood to express not relations on real numbers, but rather the corresponding relations on numerals. Thus (i)–(iii) of article I.B.3.d might be read:

(i$^\uparrow$) x is less massive than y is
(ii$^\uparrow$) X marks how massive x is
(iii$^\uparrow$) X marks less than Y does

The second step in the strategy is to introduce modality and make the following kinds of replacements. Replace primary quantifications $\exists x$ or 'there is an x' by $\exists^@ x$ or 'there actually is an x', but secondary quantifications

$\exists X$ or 'there is an X' by $\exists^{\diamond} X$ or 'there could have been an X'. Replace primitives like (i$^{\uparrow}$)–(iii$^{\uparrow}$) above by primitives like (i$^{\Uparrow}$)–(iii$^{\Uparrow}$) below:

(i$^{\Uparrow}$) x (actually) is less massive than y (actually) is

(ii$^{\Uparrow}$) X would (necessarily) have marked (if it had existed) how massive x (actually) is

(iii$^{\Uparrow}$) X would (necessarily) have marked (if it had existed) less than Y would (necessarily) have done (if it had existed)

The result is a two-sorted theory $T^{\Uparrow}$ in a two-sorted language $L^{\Uparrow}$. The language $L^{\Uparrow}$ will have for each primary primitive F of L a corresponding primitive $F^{@@}$ like (i$^{\Uparrow}$), for each mixed primitive G of L a corresponding primitive $G^{@\square}$ like (ii$^{\Uparrow}$), and for each secondary primitive H of L a corresponding primitive $H^{\square\square}$ like (iii$^{\Uparrow}$). The theory $T^{\Uparrow}$ will have for each axiom A, B, C of T a corresponding axiom $A^{\Uparrow}$, $B^{\Uparrow}$, $C^{\Uparrow}$.

What needs to be established is the metatheorem that if a formula R of L is logically implied (under standard logic) by the axioms of T, then the counterpart formula $R^{\Uparrow}$ of $L^{\Uparrow}$ will be logically implied (under the appropriate modal logic) by the axioms of $T^{\Uparrow}$. The operation $^{\Uparrow}$ will thus, besides eliminating real numbers, also preserve logical implications, and leave primary formulas essentially unchanged. In this sense, $^{\Uparrow}$ will provide a nominalistic reconstrual, and $T^{\Uparrow}$ in $L^{\Uparrow}$ will provide a nominalistic reconstruction, of T in L.

b. The Strategic Modal Logic vs. Standard Modal Logics

The kinds of modal notions required by the strategy just outlined are listed in the adjoining table.

Modal Logical Apparatus of Purely Modal Strategy

Symbol	Reading
$\exists^{@}x$	there (actually) exists an x such that . . .
$\exists^{\diamond}x$	there could have existed an x such that . . .
$F^{@@}xy$	x (actually) is thus-and-so relative to how y (actually) is
$G^{@\square}\,xy$	x (actually) is thus-and-so relative to how y (necessarily) would have been (if it had existed)
$H^{\square\square}\,xy$	x (necessarily) would have been (if it had existed) thus-and-so relative to how y (necessarily) would have been (if it had existed)

In addition, two universal-type quantifiers can be defined by taking $\forall^@$ to abbreviate $\sim\exists^@\sim$ and $\forall^\square$ to abbreviate $\sim\exists^\diamond\sim$. Note that two sorts of variables x, y, z, . . . and X, Y, Z, . . . are involved, and that the predicates and quantifiers pertain to how the former (actually) are and how the latter would (necessarily) have been (if they had existed). Some comment may be in order as to why just these and not other modal notions were used.

To commence, the kind of modal notions with which we have been concerned so far, possibility in the sense of **potentiality**, of what (is or isn't but) potentially could have been the case if the world had been other than as it is, is often called **metaphysical** modality, in contrast to epistemological or **epistemic** modality, of possibility in the sense of what (for all we know) may be the case in the world as it actually is. One version or variant of the epistemic modality is logical or **metalogical** modality, concerned with possibility in the sense of **non-contradiction**, of what could without self-contradiction be assumed to be the case. This may be understood absolutely or relative to some theory; in the latter case, being necessary amounts to being implied by the theory. A variant of logical modality is the contrast between the necessary in the sense of the analytic and the contingent in the sense of the synthetic. Corresponding to narrowly logical **implication** of the conclusion that Q by the premiss that P (which holds if the conditional assertion that if it is the case that P then it is the case that Q is a logical truth) one has the broader analytic **entailment** (which holds if the conditional is an analytic truth). Now why has the strategy employed metaphysical rather than metalogical or some related notion of modality? There are two reasons.

The first and more philosophical is that the usual explanations of metalogical modality by which people are introduced to the notion, say in introductory logic courses, seem to explain the notion in terms of what arguments, presumably in the sense of abstract linguistic expression types, there are, or else in terms of what arguments, in the sense of concrete linguistic expression tokens, there could have been—in other words, in terms either of abstracta or of metaphysical possibility. The second and more technical is that it is very difficult to make sense of quantifying into a modal context $\exists x \diamond Fx$ when the $\diamond$ is read as non-contradiction. This is because presumably $\exists x \diamond Fx$ should be true if and only if there is an object a such that $\diamond Fx$ is true of a; but while there is in standard metalogical theory a well-defined notion of what it is for a closed formula P to be non-contradictory, there is no such notion of what it is for an open formula Fx to be non-contradictory of an object a. One might try to reduce the open case to the closed by defining Fx to be non-contradictory of a if and only if Ft is non-contradictory, where t is a term denoting a. But on the one hand there may be no such term, and other the other hand there may be two such terms t', t'' with Ft' non-contradictory and Ft'' contradictory.

To continue, why was it that the possible existence of concrete numerals rather than that of abstract numbers was assumed? The reason is perhaps mainly that the idea of a distinction between actual and possible existence makes questionable sense in application to pure mathematicalia like numbers and sets of numbers (a topic discussed in article I.A.1.b); and whether or not it makes sense, the assumption that mathematical entities could perfectly well have existed and just happen not to is one few nominalists have found attractive. Indeed, even the appeal to the possible existence of new sorts of entities that are concrete in that they causally interact with each other, but that do not causally interact with actually existing sorts of entities, including human beings, might be thought nominalistically repellent—though this depends on just what the motivation for sympathy with nominalism is supposed to be. Once it is conceded that if there had been things there aren't, of such a sort as would have interacted with things of the sort there are, then it must be conceded that if there had been such things, the things there are might not have been just as they are (and some of them might not have been at all), owing to interaction with these other entities: if there had been more stars than there are, for instance, the actual planets might not have been exactly where they are, owing to the gravitational attraction of those stars. And once this is conceded, the use of notions like those tabulated above becomes almost inevitable.

The discussion of section 2 was intended to show that such notions are intuitively intelligible as part of ordinary language, and in this sense suitable to be taken as primitives of a modal logic. Ideally, what should come next would be a tabulation of assumptions involving these notions required by the strategy; an informal discussion showing that such assumptions are intuitively plausible as part of commonsense thought, and in this sense suitable to be taken as axioms of a modal logic; and a proof for the modal logic with the primitives and axioms tabulated of the metatheorem needed for the strategy, as mentioned at the end of article 3.a. Or rather, ideally, given that modal logic has been pursued by scores of researchers for decades, one would expect all this to be available already in the vast existing literature. Unfortunately, an appropriate modal logic is not available in the existing literature on modal logic, though one can be put together by careful picking and choosing, mixing and matching, bits and pieces, odds and ends, from among options in the existing literature. Realistically, what the present outline can hope to accomplish within its limited scope is just to explain roughly why the existing literature does not make available ready-made an appropriate modal logic, and roughly which bits and pieces are the ones that should be picked and chosen.

One important reason why the existing literature does not supply what is wanted in the present context is that what is wanted is a logic of metaphysical possibility, but the existing literature was originally developed to supply a logic of metalogical possibility, and still reflects its origin. It does so at both the level of basic modal logic and of quantification in modal logic, as in such standard reference works as Bull and Segerberg (1984) and Garson (1984), respectively. R. A. Bull and Krister Segerberg have to take note of an appalling array of competing modal sentential logics, differing mainly in their treatment of iterated modalities. In metalogical modal logic, especially where possibility and necessity are understood as consistency and provability not absolutely but relative to some theory, iteration of modality makes sense, and makes important differences: after all, the Incompleteness Theorem of Kurt Gödel shows there is an important distinction to be made between mere consistency and provable consistency. By contrast, in metaphysical modal logic, as was mentioned in passing in section 1, while each verb may be put in the subjunctive or conditional mood, there are no 'double subjunctive' or 'subjunctivo-conditional' or 'double conditional' moods. It seems an intuitively plausible assumption that once a modal modification has been made, further modal modifications are not permitted, or if permitted do not make any difference: the weakest further modification (possibility) and the strongest (necessity) ought to be equivalent to each other and to no further modification—an assumption called **rigidity**. The preoccupation of the existing literature with non-rigid systems is surely to a large degree a reflection of the original metalogical orientation of the subject.

James W. Garson has to take note of an appalling array of competing proposals for combining modality and quantification. Many of these involve the assumption that predicates have existential and coexistential import. Now quantification into metalogical modal contexts is difficult to make sense of except in certain very special situations (for instance, where every object *a* has a unique preferred or canonical name *t*), and it may be that in such very special situations existential and coexistential import is a reasonable assumption. By contrast, in metaphysical modal logic, as has already been mentioned in section 2, the only intuitively plausible assumption is that in general predicates do not have such import—the assumption called **freedom**. The occurrence of unfree systems in the existing literature perhaps is to some degree a reflection of the original metalogical orientation of the subject.

Another important reason why the existing literature does not supply what is wanted in the present context is that it treats possibility as an operator applying to whole formulas, whereas distinctions of mood apply to verbs. This difference makes a difference as regards both molecular and atomic formulas. As to molecular formulas, when a modality $\diamondsuit$ is applied to such a formula B, it governs every subformula of B. By contrast, in ordinary language, when the verb in the main clause of a sentence is put in the subjunctive, the verb in a subordinate clause often may be put in either the conditional or the indicative. As to atomic formulas, when a modality $\diamondsuit$ is applied to an atomic formula $F(x, y)$, it governs both places of the predicate F. By contrast, in ordinary language, predicates may, and predicates of comparison often do, contain two verbs, either of which may be put in the subjunctive or the indicative independently of the other. These difficulties and analogous ones in tense logic may be illustrated by a few examples:

(i)   When she was rich, all those who (now) scorn her (then) praised her.

(i′)   If she had been rich, all those who (now, actually) scorn her (then, consequently) would have praised her.

(ii)   When the entail is broken, the daughter will be as rich as the son (then) will be, namely, half as rich as the son (now) is.

(ii′)   If the entail had been broken, the daughter would have been as rich as the son (then, consequently) would have been, namely, half as rich as the son (now, actually) is.

(iii)   When his junk bonds pay off, he will be even richer than she will be when her junk bonds pay off.

(iii′) If his junk bonds had paid off, he would have been even richer than she would have been if her junk bonds had paid off.

(These examples incidentally illustrate the fact that the analogy between modality and temporality is so close that in ordinary language 'always' and 'now' and 'never' are often used not in their primary temporal senses but rather in secondary modal senses interchangeable with 'necessarily' and 'actually' and 'impossibly'.)

Despite all this, it is (just barely) possible to patch together an appropriate modal logic for the notions tabulated at the beginning of this section by mixing and matching odds and ends in the standard literature.

First, there is one sentential modal logic among the standard options in the existing literature of the kind surveyed by Bull and Segerberg that is compatible with the assumption of rigidity, namely the one called $S5$, with the characteristic theorems:

$$\Diamond\Box P \leftrightarrow \Box P \leftrightarrow \Box\Box P$$
$$\Diamond\Diamond P \leftrightarrow \Diamond P \leftrightarrow \Box\Diamond P$$

Second, there are, among the various options in the existing literature of the kind surveyed by Garson, approaches to adding quantifiers to one's chosen system of sentential modal logic that are compatible with the assumption of freedom, such as the approach leading to a system called $Q1R$. Adopting such a system, the quantifier $\exists^\Diamond X$ can be expressed in more conventional notation as $\Diamond\exists X$, and the primitive $H^{\Box\Box}XY$ expressed in more conventional notation as $\Box H$. (Rigidity for $H^{\Box\Box}$, the equivalence of $\Diamond H^{\Box\Box}$ and $H^{\Box\Box}$ and $\Box H^{\Box\Box}$, then becomes provable.)

Third, there are some available optional additional operators in existing literature that help with some of the problems illustrated by the examples (i′)–(iii′). The relevant literature began with work on a temporal **now** operator by Hans Kamp, and a temporal **then** operator by Frank Vlach, some information about which can be found in the survey of basic tense logic, Burgess (1984*a*). An analogous modal 'now' or **actually** operator @ has been considered. Its effect is to restore the indicative mood in subordinate clauses (in particular, it makes no difference unless it occurs within the scope of a $\Box$ or $\Diamond$). Also an analogous modal 'then' or **consequently** operator ¢ has been considered. Its effect is to restore the conditional mood in subordinate clauses (in particular, it makes no difference unless it occurs within the scope of an @). Such operators are helpful with the problem of shifting moods in subordinate clauses, as in (i′), which can be formalized as:

$$\Box(Px \rightarrow @\forall y(Qxy \rightarrow ¢Rxy))$$

A useful source of information about them is the work of Harold Hodes, especially the study Hodes (1984*a*). Adopting such a system, the quantifier $\exists^@x$ can be expressed in more conventional notation as $@\exists x$, and the primitive $F^{@@}XY$ as $@F$. (Rigidity for $F^{@@}$, the equivalence of $\Diamond F^{@@}$ and $F^{@@}$ and $\Box F^{@@}$ then becomes provable.) Still, even the new operators are not helpful where the problem is about independence of moods in the two places of a predicate, as in example (ii′) of actual/hypothetical cross-comparison and (iii′) of hypothetical/hypothetical cross-comparison. The problem is that there is no way to indicate in conventional notation the fact that in the mixed primitive $G^{@\Box}$ the first place is modally modified in one way and the second place in another.

But fourth and finally, one can at least express the fact, for instance, that in $G^{@\Box}$ each place is modally modified in some way or other. Or rather, one can express the rigidity that is assumed to follow from this fact, and which is provable for the primary and secondary primitives when they are expressed as indicated above. Namely, one can express the primitive $G^{@\Box}$ by a simple G, and just add rigidity as an axiom. While this does not express everything, in the present context (where the same variable never occurs both in some place of some predicate that is governed by 'actually' and in some place of some other predicate that is governed by 'necessarily'), it expresses enough. Enough, that is to say, to render provable the metatheorem that is needed for the strategy, as stated at the end of article 3.a, which follows immediately from another metatheorem that may be stated as follows: for any modal formula Q, let Q^∇ be the result of dropping all modalities from Q. Then for any relevant modal formulas, R is deducible from P if and only if R^∇ is deducible by P^∇. The (outline of the) proof is relegated to the following optional semitechnical appendix.

c. Sketch of the Proof

For modal logic whose sentential component is the rigid system S5, a model $\mathcal{M}$ consists of a set I of indices, one of which h is designated the **home** index. For each i in I there is a universe $\Gamma(i)$ (and for a two-sorted language another universe $\Delta(i)$) and a specification for each primitive F and each a, b, ... of whether or not $F(a, b, \ldots)$ is true at i. For a modal logic whose quantificational component is the free system Q1R, this means a specification not just for all a, b, ... in $\Gamma(i)$, but for all a, b, ... in the union Γ of all the $\Gamma(i)$ (and for a two-sorted language similarly for the union Δ of all the $\Delta(i)$).

The definition of truth at an index for molecular formulas is analogous to that for standard logic. The negation and disjunction clauses require no comment. The clauses for existential quantification, possibility, and (if present) actuality are as shown in the first four lines of the adjoining table. (For a two-sorted language there would be an additional clause for secondary existential quantification parallel to the clause shown for primary existential quantification.) Truth in the model as a whole means truth at the home index.

Model Theory for Modal Logic

Formula	Truth condition at i
$\exists x\, Q(a, b, \ldots, x)$	for some c in $\Gamma(i)$, $Q(a, b, \ldots, c)$ is true at i
$\Diamond Q(a, b, \ldots)$	for some j, $Q(a, b, \ldots)$ is true at j
$@Q(a, b, \ldots)$	$Q(a, b, \ldots)$ is true at h
$@\exists x\, Q(a, b, \ldots, x)$	for some c in $\Gamma(h)$, $Q(a, b, \ldots)$ is true at h
$\Diamond\exists x\, Q(a, b, \ldots, x)$	for some c in Γ, $Q(a, b, \ldots)$ is true at h

Note that for possibility and actuality, i is not mentioned on the right-hand side of the table, so a formula beginning with $\Diamond$ or $@$ is true at any one index if and only if it is true at any other, and in particular if and only if it is true at h. The assumption of rigidity for a primitive G amounts to the assumption that likewise for any a, b, ... it is the case that $G(a, b, \ldots)$ is true at any one index if and only if it is true at any other, and in particular if and only if it is true at h. It can now be seen that if a formula Q is built up from rigid primitives using only $\sim$ and $\vee$ and quantifiers that are immediately preceded by modalities, then it too is true at any one index if and only if it is true at any other, and in particular if and only if it is true at h. For such formulas, the clauses above boil down to what is shown in the last two lines of the adjoining table.

It can now be seen that if a formula Q is built up from rigid primitives using only $\sim$ and $\vee$ and primary existential quantifications immediately preceded by actuality $@\exists x$ and secondary quantifications immediately preceded by possibility $\Diamond\exists X$, and if $Q^{\triangledown}$ is the result of dropping all modalities from Q, then $Q(a, b, \ldots, A, B, \ldots)$ is true (at h) in the given model $\mathcal{M}$ for modal logic if and only if $Q^{\triangledown}$ is true in the model $\mathcal{M}^{\triangledown}$ for standard logic whose two universes are $\Gamma(h)$ and Δ, and for which truth for predicates is specified by taking $G(a, b, \ldots, A, B, \ldots)$ to be true in $\mathcal{M}^{\triangledown}$ if and only if it is true (at h) in $\mathcal{M}$. It follows that if R is not a consequence of P (so that there exists a model of $R \wedge \sim P$), then $R^{\triangledown}$ is not a consequence of $P^{\triangledown}$; and the converse can also (and more easily) be shown to hold. The metatheorem follows by the Completeness Theorems for modal and standard logic, according to which model-theoretic consequence and proof-theoretic deducibility coincide.

The metatheorem can also be established by a direct proof-theoretic argument.

C

A Mixed Modal Strategy

In standard textbooks of set theory, the various numbers systems of traditional mathematics are constructed from sets (along the lines sketched in article I.B.1.a). In such constructions, there are several different ways to proceed, using Zermelo's version of the natural numbers or von Neumann's, Dedekind's version of the real numbers or Cantor's. Once the construction has been carried out, however, and basic laws like the progression axioms for the usual order on the natural numbers or continuity for the usual order on the real numbers have been established, mathematicians tend to forget the details of the construction, and base subsequent proofs only on the basic laws. Hence a proof of a theorem about the natural numbers:

(o)   For the natural order on the natural numbers it is the case that . . .

typically will prove more than is stated, namely:

(i)   For any progressive order it is the case that . . .

In one of several senses of the term in contemporary mathematics, a structure is something like a set of entities, the universe of the structure, together with some distinguished relations on them, such as the set of natural numbers together with the natural order relation on them. (A structure is thus essentially the same thing as a model in the logicians' sense.) In one of several senses of the term in contemporary philosophy of mathematics (encountered already in article I.A.2.d), structuralism is, at a first approximation, the view that theorems of mathematics like (o) should be construed not as being specifically about some one structure of a given type, such as Zermelo's version of the natural numbers, or von Neumann's, but rather as being generalizations about all structures of the given type, like (i).

However, if there are no structures of the given type, it will be equally and vacuously true that:

(i′)   For any progressive order it is not the case that . . .

So structuralism is, at a second approximation, the view that theorems of mathematics like (o) should be construed as assertions like:

(ii)   For all progressive orders (and there are some)
        it is the case that . . .

In recent decades, structuralism has become, like nominalism, one of the recognized 'isms of the philosophy of mathematics, the paper Benacerraf (1965) marking the kind of watershed for the one that Benacerraf (1973) marked for the other. Usually structuralism and nominalism are considered rivals. But structuralism can also be the first step in a strategy of nominalistic reconstrual or paraphrase differing from those considered so far.

It is not in itself a very long step towards nominalistic paraphrase. For the supposition that there exists a structure of the appropriate type is, in the first place, a supposition to the effect that there exist infinitely many entities, the elements of the universe of the structure. And this is something many nominalists don't want to assume. One may, however, take a further step to a version or variant of structuralism that strengthens generalizations about all structures of appropriate type to assertions about all *possible* such structures, but weakens the supposition of the existence of such a structure to the supposition of the *possible* existence of such a structure. It thus requires only the possible existence of enough entities, which entities can be taken to be concrete. Such modalism is the second step in the strategy.

It still leaves one some way away from a nominalistic paraphrase. For the supposition that there could have existed a structure of the appropriate type is, in the second place, a supposition to the effect that there could have existed a set of infinitely many elements, the universe of the structure. (Moreover, the standard laws generally mention subsets of the universe of the structure.) The theory of one level of sets over some individuals is sometimes considered a 'logic' and so considered is called **second-order logic**, as contrasted with standard or **first-order logic**. It will not be so considered by those who understand 'logic' in such a way that a 'logic' cannot have existential implications; and whether or not it is considered a 'logic', it won't be acceptable to nominalists. Now modal structuralism assumes only the possible, not the actual, existence of a structure, and so would at worst involve only assuming the possible and not the actual existence of sets. But even this (as mentioned in article B.3.b) is something nominalists generally don't want to assume. There are, however, available two conceivable nominalistic surrogates for 'second-order logic' or the theory of one level of sets over some concrete individuals. First, there is a theory or logic of the part/whole relation introduced by the proto-nominalist Leśniewski under the name **mereology**, developed (beginning with Goodman and Leonard (1940)) by the nominalist Goodman and co-workers under the name **calculus of individuals**, and examined in all its versions and variants in Simons (1987). Second, there is a nominalistic replacement for the theory of sets of concrete entities provided by the logic or theory of **plurality** developed more recently by George Boolos (in a series of papers beginning with Boolos (1984)), which may be called **plethynticology**.

This even still does not quite bring one all the way to a nominalistic paraphrase. For the supposition that there could have existed a structure of the appropriate type is, in the third place, a supposition to the effect that there could have existed certain relations, and relations are abstracta, standardly taken in this context to be set-theoretic entities more complicated than simple subsets of the universe of the structure, namely, sets of ordered pairs. It turns out that such relations can be represented by sets of sets of elements of the universe of the structure. So nominalists who are willing to accept both mereology and plethynticology will find in the combination an appropriate nominalistic surrogate, completing the strategy. The power of the combination of mereology with plethynticology first became apparent in the work of David Lewis, in his book Lewis (1991), the contributions of Allen Hazen to the co-authored Appendix thereto, and in the follow-up article Lewis (1993). (These are among the best sources of information for the relevant aspects of mereology and plethynticology, and the only good sources for the combination.)

The strategy just sketched is plainly a very mixed modal strategy, in the sense that it combines modal logic with other extensions of standard logic. A somewhat modified version of the strategy just sketched, in which version modality is the last ingredient introduced, will be presented in this chapter. Section 1 deals with logical preliminaries, summarizing second-order logic and the nominalistic partial substitutes for it provided by plethynticology and mereology. Section 2 and article 3.a will summarize the non-modal steps of the strategy. Article 3.b will summarize the modal logic involved, which is similar but not identical to that involved in the purely modal strategy of the preceding chapter.

1. THE LOGIC(S) OF THE STRATEGY

a. Second-Order Logic

The general theory of all sets X, Y, Z, ... of basic entities x, y, z, ... is often called **monadic** second-order logic. The most usual presentation involves two fundamental assumptions. The first, **extensionality**, can be expressed in a single axiom:

(i)   for all X and Y, if for all z, $z \in X$ if and only if $z \in Y$, then $X = Y$

The second, **comprehension**, can be expressed in a scheme:

(ii)   there exists an X such that for all x, one has $x \in X$
       if and only if $F(x)$

Here $F(x)$ may be a disjunction of quite disparate clauses, so that (ii) goes a long way towards expressing the assumption that quite disparate collections of basic entities can be sets. (The most usual version of the theory actually has a third fundamental assumption (iii) of **choice**, the details of whose formulation will be suppressed here, which goes even further in this direction than the assumption (ii) of comprehension.)

The theory can be generalized in either of two directions. Generalizing in one direction, there is an analogous theory of two-place or binary relations X, $\mathcal{Y}$, Z, ... on basic entities x, y, z, ..., called **dyadic** second-order logic, again with assumptions of extensionality and comprehension (and choice). Writing u; $v \ni X$ for 'u stands in relation X to v', extensionality and comprehension can be expressed by exact analogues of (i), (ii), namely:

(i*)   for all X and $\mathcal{Y}$, if for all u and v, one has u; $v \ni X$
       if and only if u; $v \ni \mathcal{Y}$, then $X = \mathcal{Y}$
(ii*)   there exists an X such that for all u and v, one has u; $v \ni X$
       if and only if $F(u, v)$

Similarly one may formulate triadic, tetradic, ... second-order logic. Full or **polyadic** second-order logic includes the monadic, dyadic, triadic, ... assumptions.

Generalizing in another direction, there is an analogous theory of sets Ξ, Υ, ... of sets X, Y, ... of basic entities x, y, ..., called monadic **third-order logic**, again with assumptions of extensionality and comprehension (and choice). Writing $X\varepsilon\Xi$ for elementhood of a set in a set of

sets, extensionality and comprehension can be expressed by exact analogues (i#), (ii#) of (i), (ii). Beyond this there are fourth-, fifth-, . . . order
logics. Full (monadic) **higher-order logic** includes the assumptions
of (monadic) second-, third-, fourth-, . . . order logic. (It amounts to the
simplified theory of types encountered in article I.B.1.b.)

One could also consider generalizing in both directions simultaneously
to obtain a polyadic higher-order logic; but this turns out to add nothing
essentially new. That is because two-place relations X can be represented
by sets of ordered pairs (u, v), and the ordered pair (u, v) can be represented as a set of sets $\{\{u\}, \{u, v\}\}$, and similarly for ordered triples,
quadruples, . . . Note, however, that with this way of proceeding, dyadic
second-order logic is reinterpreted in monadic fourth-order logic, dyadic
third-order logic in monadic fifth-order logic, and so on. A trick (alluded
to towards the end of section o) permits some improvement here, namely,
the reinterpretation of dyadic second-order logic in monadic third-order
logic. For while in general a relation is taken to be a set of ordered pairs
$(u, v) = \{\{u\}, \{u, v\}\}$ and hence a set of sets of sets, in special cases a
relation can be represented by a set of sets. For instance, an order relation
$<$ can be represented by the set of all its **segments** or sets of form $\{u \mid u < v\}$. For another instance, a symmetric relation can be represented by
the set of all unordered pairs $\{u, v\}$ such that u is related to v. Finally,
an arbitrary relation X can be represented, not indeed by a single set of
sets, but by a quadruple of them, $\Upsilon, \Xi^<, \Xi^\circ, \Xi^>$, where:

Υ represents an order relation $<$ on the basic entities
$\Xi^<$ $= \{\{u, v\} \mid u < v \text{ and } u; v \ni X\}$
Ξ° $= \{\{u\} \mid u; u \ni X\}$
$\Xi^>$ $= \{\{u, v\} \mid v < u \text{ and } u; v \ni X\}$

(Here choice guarantees the existence of a suitable $<$.)

Returning now to monadic second-order logic, it admits of several
versions or variants besides the formulation considered above. There is a
minor variant that may be described as follows. The formulation considered above supposes that the logical predicate of identity $X = Y$ is
taken as a primitive, with the logical axiom and scheme:

(0.α)   for all X, $X = X$
(0.β)   for all X and Y, if $X = Y$, then $G(X)$ if and only if $G(Y)$

But alternatively, (i) could be taken as the definition of identity, in which
case (0.α) would be deducible, and in order to render (0.β) deducible,
extensionality would be assumed in the form of a scheme:

(i′)   for all X and Y, if for all z, $z \in X$ if and only if $z \in Y$,
        then $G(X)$ if and only if $G(Y)$

There is a more substantial but still not major variant that avoids the
assumption of a **null** set $\{\ \}$ with no elements, and of the **unit** set $\{x\}$
with just the element x as something over and above x itself, by making
certain tedious but routine changes in the formulation of comprehension
(and choice). The theory with the null set and unit sets is easily
reinterpretable within the theory without them (so that philosophers who
strain at the gnat of a null set or a unit set while swallowing the camel of
arbitrary sets, so long as the latter have more than one element, can
regard talk of the null and unit sets as a dispensable convenience, a
harmless manner of speaking). Essentially this is because, letting X, Y,
Z, ... range over all sets, and X', Y', Z', ... over sets with more than
one element:

(iv)   there exists an X such that $G(X)$

is equivalent to:

(v) $G(\{\ \})$ or there is an x such that $G(\{x\})$ or there is an X' such
        that $G(X')$

and hence to:

(vi) G' or there is an x such that $G''(x)$ or there is an X' such that
        $G(X')$

where G' is the result of replacing each $y \in X$ in $G(X)$ by $y \neq y$, and
$G''(x)$ is the result of replacing each $y \in X$ in $G(X)$ by $y = x$.

Finally, there is a major variant. Given a set I of basic entities
x, y, z, ..., the theory of all subsets X, Y, Z, ... of I can be formulated
in a way that avoids any explicit mention of the basic entities x, y, z, ...:
there are variables X, Y, Z, ... for subsets of I, and the predicate $X \subseteq Y$
for inclusion. Explicit mention of the basic entities x, y, z ... can be
avoided essentially because such an entity x can be represented by its unit
set $\{x\}$, with $x \in Y$ holding if and only if $\{x\} \subseteq Y$; and because it is
possible to characterize which X are unit sets purely in terms of $\subseteq$: define
X to be **void** if there is no Z other than X itself such that $Z \subseteq X$, and
to be **atomic** if it is non-void and there is no non-void Z other than X
itself such that $Z \subseteq X$. Then a set is void if and only if it is the null set
and a set is a unit set if and only if it is atomic.

The standard $\subseteq$ axioms for sets X, Y, ... of entities not themselves explicitly mentioned are those known to mathematicians as the axioms for a **complete atomic Boolean algebra** (plus choice). The Boolean axioms admit of many equivalent formulations well known to mathematicians. In what is perhaps the most convenient formulation, the list of **Boolean** axioms begins with the **partial order** axiom, the conjunction of the three assumptions of **reflexivity** and **anti-symmetry** and **transitivity**:

$$X \subseteq X$$
$$\text{if } X \subseteq Y \text{ and } Y \subseteq X, \text{ then } X = Y$$
$$\text{if } X \subseteq Y \text{ and } Y \subseteq Z, \text{ then } X \subseteq Z$$

Anti-symmetry (much as with extensionality) can also be formulated as a scheme:

$$\text{if } X \subseteq Y \text{ and } Y \subseteq X, \text{ then if } G(X), \text{ then } G(Y)$$

The list of Boolean axioms would continue with those that guarantee the existence of certain sets characterized by certain inclusion relationships with other sets, namely, those listed in the adjoining table.

Characterization of Set-Theoretic Notions in Terms of Inclusion

Symbol	Name	Characterization
ø	Null Set	for all Z, $\varnothing \subseteq Z$
I	Universal Set	for all Z, $Z \subseteq I$
$X \cap Y$	Intersection	for all Z, $Z \subseteq X$ and $Z \subseteq Y$ if and only if $Z \subseteq X \cap Y$
$X \cup Y$	Union	for all Z, $X \subseteq Z$ and $Y \subseteq Z$ if and only if $X \cup Y \subseteq Z$
$-X$	Complement	$X \cap -X = \varnothing$ and $X \cup -X = I$
$X - Y$	Difference	$X \cap -Y$
X/Y	Quotient	$X \cup -Y$

(Related to the notions in the table is the following: X, Y are called **disjoint** or **overlapping** according as $X \cap Y$ is or is not null.) The list of Boolean axioms would conclude with some algebraic axioms involving the operations of intersection, union, and complementation, the details of whose formulation will be suppressed here.

The notions of intersection and union can be defined more generally. Y is the intersection of all X such that $G(X)$ if and only if:

$$\text{for all } Z, Z \subseteq Y \text{ if and only if } Z \subseteq X \text{ for all } X \text{ such that } G(X)$$

Y is the union of all X such that $G(X)$ if and only if:

> for all Z, $Y \subseteq Z$ if and only if $X \subseteq Z$ for all X such that $G(X)$

The assumption of **atomicity** is that every X is the union of all atomic sets included in it. The assumption of **completeness** is that the union of all X such that $F(X)$ exists, even where (much as with comprehension) $F(X)$ may be a disjunction of quite disparate clauses.

b. Plethynticology

The simplest distinctions of multiplicity are expressed in natural languages like English through 'number' (in the grammatical sense), the distinction between singular and plural, a feature of the system of declension of nouns. More complicated distinctions are expressed using special nouns, such as 'collection' in everyday contexts, or 'set' or 'class' in mathematical ones. Generally, what can be expressed using plurals, for instance:

(o)   There are some critics who admire only each other

in the sense of:

(i)   There some critics such that a critic who is one of them
        admires another critic if and only if
            that other critic is also one of them

can be regimented using 'set' or one of its cognates:

(ii)   There is a set of critics such that a critic in the set
        admires another critic if and only if
            that other critic is also in the set

But not everything that can be expressed using set-theoretic language can be expressed using plurals: very roughly, singular assertions about sets of basic entities can be paraphrased as plural assertions about the basic entities, and singular assertions about sets of sets of basic entities can be paraphrased as plural assertions about sets of basic entities, but singular assertions about sets of sets of basic entities cannot be paraphrased as plural assertions about basic entities.

In mathematics, since the beginning of the nineteenth century, there has been an ascent to higher and higher levels of abstraction: first there was differential and integral calculus, a theory of real numbers and certain especially important sets of or functions on real numbers, such as the exponential function. Next came real analysis, a general theory of sets of or functions on the real numbers, and of certain especially important functionals, or functions on sets or functions, such as the definite integral functional. Last came functional analysis, a general theory of sets of or functions on sets of or functions on real numbers. Correspondingly, a sophisticated ultra-general theory of higher and higher levels of sets or classes (and functions or operations) has come to be taken as a framework for mathematics (as sketched in article I.B.1.a).

Modern logic was initially developed for purposes of analysing mathematical arguments. Hence it is perhaps not surprising that no provision was made in it for plurals: if one wishes to apply it to arguments turning on 'number' (in the grammatical sense), the simplest procedure seems to be simply to regiment plural assertions about the Fs into singular assertions about the set of all Fs. Such regimentation has seemed artificial in a derogatory sense to some, notably Boolos, who in the work cited in section o pioneered the development of an autonomous plural quantification logic or plethynticology.

This logic enriches standard logic with the apparatus shown in the adjoining table. In addition the dual quantifier $\forall\forall xx$, defined as $\sim\exists\exists\, xx\sim$ may be used.

Logical Apparatus of Plethynticology

Symbol	Reading
$\exists\exists xx$	there are some things, the xs, such that . . .
$y == xx$	y is one of the xs

The 'syntax' or proof theory of plethynticology involves three schemes: First, there is **extensionality**:

$$\forall xx \forall yy (\forall z (z == xx \leftrightarrow z == yy) \rightarrow (Q(xx) \leftrightarrow Q(yy)))$$

If anything is one of the xs if and only if it is one of the ys, then (the xs are identically the ys and hence) anything that holds of the xs holds of the ys. Second, there is **comprehension**, which amounts to:

$$\exists xx \forall y (y == xx \leftrightarrow Q(y))$$

There are some things such that anything is one of them if and only if Q holds of it. (Which things? Those things such that Q holds of them!) Third, there is **choice**, the details of whose formulation will be suppressed here.

These correspond to the axioms of second-order logic. More specifically they correspond to the minor variant of the most usual axioms that avoids taking identity between sets as primitive. This version of plethynticology reads 'there are some . . .' as meaning 'there are zero or more . . .' rather than 'there are one or more . . .' or 'there are two or more . . .'; but the variations in the axioms needed to accommodate one of the other readings would not be major. They would be analogous to the modifications in the axioms of second-order logic needed to avoid the null set or unit sets.

The customary rules of inference for plural existential quantification and singular-plural identity parallel those for (singular) existential quantification and (singular-singular) identity in standard logic, so that the following metatheorem holds: for any formula Q of plethynticology with singular and plural variables x, y, . . . and xx, yy, . . . for basic entities, let Q^V be the formula of second-order logic with variables x, y, . . . and X, Y, . . . for basic entities and sets thereof that results from replacing plural quantifications $\exists\exists xx$ by set quantifications $\exists X$, and singular-plural identity formulas $y == xx$ by $y \in X$. Then for any P and R, R is deducible by plethynticology from P if and only if R^V is deducible by second-order logic from P^V. It is in this sense that plethynticology can provide a surrogate for some uses of set theory.

The 'semantics' in the logicians' sense, the model theory, for plethynticology, reintroduces sets. Thus the models for the plethynticological formalization of (i) above:

(iii) $\exists\exists xx \forall y(y == xx \rightarrow \forall z(Ayz \leftrightarrow z == xx))$

would be taken to be just the models for the formalization of (ii) above:

(iv) $\exists X \forall y(y \in X \rightarrow \forall z(Ayz \leftrightarrow z \in X))$

Philosophical issues about the propriety of the use of plethynticology by professed nominalists resemble those about modality (as discussed in Chapter B): in particular, there is a hermeneutic reductivism about plurality, expressed for instance in Resnik (1988), that maintains that the 'semantics' in the linguists' sense of the meaning theory for plurals should also involve sets, that there is a hidden involvement with collections of Fs in talking in the plural of Fs.

Though this issue (which closely parallels the corresponding one about modality) will not be extensively discussed here, it may be remarked that in Boolos (1984) an example is given that seems an exception to the generalization that whatever can be said using plurals can be said set-theoretically. Boolos observes that it seems true that:

> there are some sets such that any set is one of them if and only if it is not an element of itself

though it is demonstrably false—the famous Russell paradox—that:

> there is a set such that any set is an element of it if and only if it is not an element of itself

c. Mereology

Mereology is a theory of the part/whole relation, here to be symbolized $\propto$. In its most convenient formulation, the axioms of mereology are those known to mathematicians as the axioms for a **complete Boolean algebra** (plus choice). They are thus the axioms for set-theoretic inclusion $\subseteq$ in the theory of subsets X, Y, ... of some given set, except that one writes $y \propto x$ rather than $Y \subseteq X$, and drops the assumption of atomicity. In its most customary formulation, the axioms of mereology are the minor variants of these needed to avoid assuming the existence of a null entity. Various (equivalent) axiomatizations are considered in detail in Simons (1987: chapters 1, 2).

Mereology is sometimes considered a 'logic', providing a logical theory of the part/whole relation much as classical logic provides a logical theory of the identity relation. It will not be so considered by those who understand 'logic' in such a way that a 'logic' cannot have existential implications. For mereology does have existential implications. If it is accepted, then the acceptance of some initial entities involves the acceptance of many further entities, arbitrary wholes having the initial entities as parts. Notably, acceptance of arbitrary material bodies (and parts thereof) and of mereology involves the further acceptance of conglomerates, or arbitrary unions of different bodies (or parts thereof); also, acceptance of geometric points and of mereology involves the further acceptance of geometric regions. In this respect mereological theory resembles set theory, acceptance of which together with the acceptance of some initial entities involves the acceptance of many further entities, the sets having the initial entities as elements. Whether or not mereology is considered a 'logic', the propriety of its use by professed nominalists, and in particular the nominalistic acceptability of conglomerates composed of bits and pieces, odds and ends, from different bodies, is controversial (as mentioned in section I.A.0; see also Simons (1987)).

If it is accepted, mereology provides a surrogate for some uses of set theory, owing to the analogy between the axioms for $\propto$ and those for $\subseteq$. (According to Lewis (1991), there is more than analogy at work here: the set-theoretic relation of inclusion is literally what mereological relation of parthood amounts to for sets.)

The analogy is imperfect inasmuch as mereology does not assume atomicity. However, if the U such that $F(U)$ are disjoint, and if the variables X, Y, Z, ... are restricted to range only over entities that are unions of such U, then atomicity will hold and the U such that $F(U)$ will count as atomic in the sense that for such a U there will be no non-null V other than U itself among the entities over which the variables range that is a part of U.

2. VARIANTS OF ANALYSIS

a. The Original

There are several standard variants of the version of analysis we have been using so far. These have differing sorts of variables for different sorts of entities, differing primitives, and differing axioms, as listed in the adjoining tables.

Recall that the original version of analysis (as in article I.B.1.b) had variables for real numbers; primitives for sum, product, order, and integrity on real numbers; and algebraic axioms for sum, product, and order, plus the continuity axiom (scheme) for order, and appropriate axioms for integrity.

The standard set-theoretic construction of the real numbers system (outlined in article I.B.1.a) can be run in reverse, so to speak. When this is done one obtains successive reductions or transformations of the original version of analysis to others with different apparatus. These will be considered in the articles immediately following. Then some further set-theoretic transformations will be considered in the articles following that, in preparation for the introduction of plethynticology and mereology into the formulation of analysis in the last articles of this section. The series of transformations will be resumed at the beginning of article 3.b, where an appropriate modal logic is introduced.

Version	Variables
a	real numbers X
b	rational numbers ξ
b	sets of rational numbers X
c, d, e	natural numbers ξ
c, d, e, f, g	sets of natural numbers X
d	binary relations on naturals X
e, f, g	sets of sets of natural numbers Ξ
h, i, j, k, l	sets of basic entities of an unspecified sort X
h, i, j	sets of sets of basic entities of an unspecified sort Ξ
m, n	(complex) basic entities of an unspecified sort X
p	(extended) physical entities of an otherwise unspecified sort X

Version	Primitives (non-set-theoretic)
a	real sum, product
a	real order
a	integrity
b	rational sum, product
b	rational order
c	natural sum, product
c, d, e	natural order
f	the analogue for unit sets of naturals of usual order on naturals
g	segmenthood
h, i	distinction

Version	Set-theoretic primitives
b	elementhood of a rational in a set of rationals
c, d	elementhood of a natural in a set of naturals
d	relatedness of a pair of naturals by a binary relation on naturals
e, f, g	elementhood of a set of naturals in a set of sets of naturals
f, g	inclusion for sets of naturals
h, i, j, k	inclusion for sets of basic entities
h, i, j	elementhood of a set of basic entities in a set of sets of basic entities

Version	Axioms (non-set-theoretic)
a	algebraic axioms for real addition, multiplication
a	algebraic axioms for real order
a	continuity axiom for real order
a	appropriate axioms for integrity
b	arithmetic for rational addition, multiplication
b	arithmetic axioms for rational order
c	arithmetic axioms for natural addition, multiplication
c, d	progression axioms for order on naturals
e, f	progressive order axioms for analogue for unit sets of naturals of usual order on naturals
g	the segments are nested and generative
h, i	the distinguished sets are nested
h	the distinguished sets are generative
j	existence of a nested set of sets of basic entities
k, l	existence of some nested sets of basic entities
m	existence of some nested basic entities
n	existence of infinitely many disjoint basic entities
p	existence of some infinitely many spatially disjoint physical entities

Version	Set-theoretic axioms
b	standard axioms for elementhood of a rational in a set of rationals
c, d	standard axioms for elementhood of a natural in a set of naturals
c	analogues of standard axioms for relatedness of naturals by a binary relation on naturals
c, d, e, f, g	standard axioms for elementhood of a set of naturals in a set of sets of naturals
f	the atomicity axiom for inclusion for sets of naturals
f, g	standard axioms for inclusion for sets of naturals, except atomicity
h, i, j	standard axioms for elementhood of a set of basic entities in a set of sets of basic entities
h, i, j, k	standard axioms for inclusion for sets of basic entities, except atomicity

The final result will be a version of analysis with no non-logical primitives or axioms of its own—in other words, a reduction of analysis to logic. Such a reduction was the goal of the logicist programme of Frege, Russell, and Ramsey; but the kind of logic they had in mind was something very different from modal plethyntico-mereology!

b. From Real to Rational

First, undoing the construction of the reals from the rationals, there is a version with the following apparatus: variables for rationals and sets thereof; primitives for sum, product, order on rational numbers, and for elementhood ∈; axioms for sum, product, order on the rational numbers; and the standard axioms for elementhood (as in article 1.a).

c. From Rational to Natural

Second, undoing the construction of the rationals from the naturals, there is a version with the following apparatus: variables for natural numbers and sets thereof; primitives for sum, product, order on natural numbers, and for elementhood; arithmetic axioms for sum, product, and the progression axioms for order, and the standard axioms for elementhood.

d. From Arithmetic to Order

Third, undoing the construction of arithmetic from order, there is a version with the following apparatus: variables for natural numbers, sets thereof, and two-place relations thereon; primitives for order, for elementhood, and for its dyadic analogue; and progression axioms for order, and the standard axioms for elementhood and their dyadic analogues.

e. From Relations to Sets of Sets

We now apply the reduction (alluded to in section o and explained in article 1.a) of dyadic second-order logic to monadic third-order logic. The resulting new version has the following apparatus: variables for natural numbers, sets thereof, and sets of sets thereof; primitives for order, elementhood between numbers and sets, and higher-order elementhood between sets and sets of sets; the progression axioms for order, and the standard axioms for elementhood and higher-order elementhood.

f. From Elementhood to Inclusion

Given sets of numbers, the explicit mention of natural numbers is super-fluous, since a number ξ can always be represented by its unit set $\{\xi\}$, and the order relation $\leq$ on numbers by the analogous relation on their unit sets, and so on. The preceding version of analysis can be reduced to a version with the following apparatus: variables for sets of numbers and for sets of sets of numbers; primitives for the analogue $\angle$ for unit sets of the order relation on numbers, for inclusion of one set of numbers in another, for (higher-order) elementhood of a set in a set of sets; the progression axioms for $\angle$, the standard axioms for inclusion (as in article 1.a), and the standard axioms for (higher-order) elementhood.

g. From Unit Sets to Segments

In some respects it would be more convenient to take a natural number ξ to be represented not by its unit set $\{\xi\}$ but by the segment in the standard order that it determines $\{\upsilon \mid \upsilon < \xi\}$. A new primitive $\int$ distinguishing the segments will be needed, since segmenthood cannot be defined in terms of inclusion; but the primitive $\angle$ will no longer be needed, since the relation on segments corresponding to the order relation on numbers is just inclusion. Sets progressively ordered by inclusion may be said to be **nested**, sets such that every set (among those over which the variables range) is a union of differences of them may be called **generative**. Since a set is a unit set if and only if it is a difference between some segment and the segment that is its immediate successor in the inclusion order, the atomicity axiom that every set (among those over which the variables range) is a union of unit sets may be replaced by the axiom that the segments are generative.

This version has the following apparatus: variables for sets of natural numbers and sets of sets of natural numbers; primitives for segmenthood, and for inclusion and (higher-order) elementhood; the axiom that the segments are nested and generative, the standard axioms for inclusion except atomicity, and the standard axioms for elementhood.

h. Towards a Structuralist Variant

The next variant involves no change in the formal theory, but merely in the intuitive understanding of it. It is a step in the direction of structuralism, namely, to drop the assumption that one is dealing with sets (and sets of sets) specifically of numbers, along with the assumption that the W such that $\int(W)$ are distinguished specifically by being segments. Instead, it is merely assumed that one is dealing with sets of basic entities of some sort and sets of sets of basic entities of that sort, and that the W such that $\int(W)$ are somehow distinguished.

The apparatus in this version is the following: variables for sets and for sets of sets of basic entities; primitives for being distinguished, for inclusion, and for (higher-order) elementhood; the axiom that the distinguished sets are nested and generative, and the standard axioms for inclusion except atomicity, and for (higher-order) elementhood.

i. A Relativized Variant

The next variant drops the axiom of generativity. The version with generativity can be reduced to a version without, simply by reinterpreting assertions about 'all X . . .' as assertions about 'all X that are unions of differences of distinguished sets of basic entities'.

j. A Structuralist Variant

The next variant takes a further step in the direction of structuralism, dropping the designation of some one, specific set of nested sets of basic entities as distinguished, weakening the axiom that the W such that $\int(W)$ are nested to the axiom that some set of sets of basic entities is so, and strengthening theorems about the specific nested sets of those W such that $\int(W)$ to theorems about *all* sets of sets of basic entities that are nested.

This leaves one with the following apparatus: variables for sets and for sets of sets of basic entities; primitives for inclusion and (higher-order) elementhood; the axiom that there exists a set of sets of basic entities that is nested, and the standard axioms for inclusion other than atomicity, and the standard axioms for (higher-order) elementhood.

k. A Plethynticological Variant

The next step is to adopt plethynticology. Then singular quantifiers $\exists\Xi$ over sets of sets may be replaced by plural quantifiers $\exists\exists XX$ over sets of basic entities, and the elementhood primitive $Y\varepsilon\Xi$ by the singular-plural identity primitive $Y == XX$. Then the axioms for ε become the axioms of plethynticology, and may be counted as belonging not to analysis in particular, but to logic in general.

The non-logical apparatus of analysis then becomes: variables for sets of basic entities; a primitive for inclusion; the axiom of the existence of some sets of basic entities that are nested, and the standard inclusion axioms except atomicity.

l. A Mereological Variant

The next step is to adopt mereology. Then $\subseteq$ may be replaced by $\propto$. The axioms for $\subseteq$ become the axioms of mereology, and may be counted as belonging not to analysis in particular, but to logic in general.

m. An Even More Structuralist Variant

The next variant takes yet another step in the direction of structuralism. It involves no change in the formal theory, but merely in the intuitive understanding of it. Namely, the understanding that X, Y, . . . are specifically sets of more basic entities may be dropped. The X, Y, . . . are just entities of some sort with the property that the mereological fusion of any entities of that sort is again of that sort, so that the entities in question are in general complex (having parts) rather than simple (or atomic, having no parts), but otherwise need not be specified.

The non-logical apparatus of analysis then becomes: variables for these entities, which now may as well be considered 'basic'; no non-logical primitives; the axiom of the existence of some nested basic enitities.

n. From Nesting to Infinity

The existence of nested entities

$$ W_0 \quad \propto \quad W_1 \quad \propto \quad W_2 \quad \propto \quad W_3 \quad \propto \quad W_4 \ldots $$

implies the existence of infinitely many disjoint entities, the differences of successive W_i. Conversely, the existence of infinitely many disjoint entities V_0, V_1, V_2, V_3, V_4, . . . implies the existence of nested entities, namely the fusions of the the Vs out to V_k (for $k = 0, 1, 2, 3, 4, \ldots$). So all that is required for the reduction of analysis is an infinity of disjuncta of the smallest, countable infinite size.

(In Lewis (1993) it is shown that within the framework of plethynticology and mereology the assumption of larger and larger uncountable infinities

of disjuncta enables one to reduce or reconstrue larger and larger portions
of standard mathematics.)

p. *A Somewhat Less Structuralist Variant*

The next variant involves no change in the formal theory, but merely in
the intuitive understanding of it, but is one that takes a step away from
structuralism. Namely, X, Y, . . . will be understood to be not just (complex) entities of some unspecified sort, but rather (extended) *physical*
entities of some otherwise unspecified sort, for which parthood amounts
to *spatial* parthood.

3. THE STRATEGY

a. *Non-Modal Aspects*

Any of the variants of analysis considered in section 2 can replace the
original version as the mathematical apparatus of an analytically formulated scientific theory (as in Chapter I.B). The form taken by mixed
primitives and axioms will have to be adjusted accordingly. In the framework of Chapter I.B, a typical mixed primitive of a scientific theory was
taken to be something like the following:

X is the mass of x (on some fixed, but here unspecified, scale)

A very simple example of an axiom involving this primitive would be:

for every x there is a real number X such that X is the mass of x

In Chapter A of this part, it was found convenient to take this notion of
absolute mass not as primitive but rather as defined in terms of an
apparatus of benchmarks and the notion of relative mass:

X is the ratio of the mass of x to the mass of y

The simple axiom would then correspondingly take the form:

for every x and y there is a real number X such that
X is the ratio of the mass of x to the mass of y

In the present chapter, it will be convenient to take this notion of exact
mass not as primitive but rather as defined in terms of the apparatus of
analysis and the notion of approximate mass:

(i.α)   the ratio of X to Y approximates the ratio of the mass of x to the mass of y

The axiom would then correspondingly take the form:

(ii.α)   for every x and y there are real numbers X and Y such that the ratio of X to Y approximates the ratio of the mass of x to the mass of y

Corresponding to each of articles 2.b–2.p there may be further changes in the form of a typical mixed primitive. The required changes are listed below:

(i.β)   the ratio of the rational number ξ to the rational number υ approximates the ratio of the mass of x to the mass of y

(i.γ,δ,ε)   the ratio of the natural number ξ to the natural number υ approximates the ratio of the mass of x to the mass of y

which may be suggestively reworded as:

        ξ stands to υ in the standard order $\leq$ on the natural numbers approximately as x stands masswise to y

(i.F)   [X, Y are unit sets of numbers and]
         X stands to Y in the order $\angle$
         approximately as x stands masswise to y

(i.ζ)   [X, Y are segments and]
         X stands to Y in the inclusion order $\subseteq$ on segments
         approximately as x stands masswise to y

(i.η,θ,ι)   [the elements of Ξ are nested and X, $Y \in \Xi$ and]
         X stands to Y in the inclusion order $\subseteq$ on the elements of Ξ
         approximately as x stands masswise to y

(i.κ,λ)   [the Ws are nested and X, Y are among the Ws and]
         X stands to Y in the inclusion order $\subseteq$ on the Ws
         approximately as x stands masswise to y

(i.μ,ν)   [the Ws are nested and X, Y are among the Ws and]
         X stands to Y in the parthood order on the Ws
         approximately as x stands masswise to y

(i.π)   [the Ws are nested and X, Y are among the Ws and]
         X stands to Y in the spatial parthood order on the Ws
         approximately as x stands masswise to y

This last will serve as the final form of the primitive originally formulated as (i.α). As the apparatus of analysis is reduced to one thing after another and the mixed primitives change from one form to another, a mixed axiom, which involves both apparatus of analysis and mixed primitives, will correspondingly change also. The final form of the axiom originally formulated as (ii.α) will be something like:

(ii.π)   for any (extended) physical Ws that are spatially nested and for any x and y there are X and Y among the Ws such that X stands to Y in the spatial parthood order on the Ws approximately as x stands masswise to y

b. Modal Aspects

The theory having been adjusted to the revised version of analysis, it takes the form of a theory T in a two-sorted language L with variables $x, y, \ldots$ for primary entities, of some physical sort, which might be called **ponderables**, and both singular $X, Y, \ldots$ and plural $XX, YY, \ldots$ variables for secondary entities, also of some physical sort, which might be called **counters**. L has some primary primitives F like:

(i)   ponderable x is less massive than ponderable y is

some mixed primitives G like:

(ii)   counter X stands to counter Y in the spatial parthood order on the Ws
         approximately as ponderable x stands masswise to ponderable y

but has no secondary primitives H except the logical primitives:

(iii)   counter X is a spatial part of counter Y
(iv)   counter X is one of the counters, the Ys

T has some primary axioms A, and some mixed axioms B each of the form:

(v) $\forall\forall WW(WW$ are nested $\rightarrow Q(WW))$

(like (ii.π) at the end of the preceding article), and the sole secondary axiom:

(vi) $\exists\exists WW(WW$ are infinitely many $\wedge WW$ are disjoint)

It will be instructive to compare this theory T to the theory of numerals from article B.3.a, the theory arrived at just prior to the introduction of modality in the strategy of the preceding chapter. To make the comparison, consider any (non-empty) disjoint entities:

$$V_0 \quad V_1 \quad V_2 \quad V_3 \quad V_4 \ldots$$

Any X that is a union of (some but not others from among) these entities can be construed as coding an infinite binary or zero–one sequence, with a one or a zero in the kth place according as V_k is or is not a part of X. Thus any such X may be construed as constituting a numeral in the very general sense of article B.3.a. Using this observation it can be shown that the theory of numerals can be reduced to the theory T under discussion. The differences between the two are that the former has heavy ideological commitments in the sense of having many non-logical primitives, while the latter has no such primitives, but has heavy logical commitments, involving as it does plethyntico-mereology.

It remains to describe the concluding step of the strategy, when modality is introduced, and the changes that are then made in (i)–(vi) above. Commencing the description with (vi), the sole axiom of analysis, asserting that there are infinitely many disjuncta, it will be weakened to the assumption that there could have been:

(vi∗) $\lozenge \exists\exists WW(WW$ are infinitely many $\wedge$ WW are disjoint)

This might almost be taken to be an axiom of logic, in which case analysis would be left without any non-logical primitives or axioms.

Given the connection between nest theory and numeral theory indicated above, (vi∗) could also be informally worded as:

(vi#)   there could have coexisted numerals for all real numbers

What the strategy of the preceding chapter assumed was something less than this:

(vi⁻)   for any given real number there could have existed a numeral for it

The syndynatontological commitments, or assumptions about what could have coexisted, of the strategy of the present chapter are heavier than those of the strategy of the preceding chapter. The modal logical commitments will in a sense be correspondingly lighter. It is precisely because the previous strategy assumed only (vi⁻) and not (vi#) that it was obliged to make constant cross-comparisons between how things actually are and how they would have been on different counterfactual hypotheses, and between how things would have been on one counterfactual hypothesis and how they would have been on another. The present strategy will only involve comparison between how things now actually are and how on the single counterfactual hypothesis of the existence of a nest they would then consequently have been.

Continuing the description of the concluding step of the strategy, a typical mixed axiom B of form (v), saying something about what all nests are like, will be replaced by a mixed axiom $B\star$ of the following form, saying something about what any nest would have been like if there had been any nests:

(v*) $\Box \forall \forall WW(WW$ nested $\rightarrow Q\star(WW))$

And what now remains to be done is to describe the changes to be made in the Q of (v) in order to obtain the $Q\star$ of (v*). The transformation $\star$ will replace primary quantifications $\exists x$ or 'there is an x' by $\exists^{@}x$ or 'there actually is an x', but secondary quantifications $\exists X$ and $\exists\exists XX$ or 'there is an X' or 'there are some Xs' by $\exists^{\mathfrak{e}}X$ and $\exists\exists^{\mathfrak{e}}XX$ or 'there would (then) have been an X' or 'there would then have been some Xs'. And it will replace primitives like (ii)–(iv) by primitives like (ii*)–(iv*) below:

(iv*) X would (then) have been one of the Ys
(iii*) X would (then) have been part of Y
(ii*) X would (then) have stood to Y
         in the spatial parthood order on the Ws
         approximately as x actually stands masswise to y

(The 'then' in every case refers back to the antecedent 'if there had been some nested Ws . . .'.) Thus the ultimate form of the simple mixed axiom taken as an example in article 3.a would be something like:

> if there had been some nested Ws then
> for every x and y that there actually are
> there would then have been X and Y among the Ws such that
> X would have stood to Y in the spatial parthood order on the Ws
> approximately as x actually stands masswise to y

The kinds of modal notions required by the strategy just outlined are listed in the adjoining table.

Modal Logical Apparatus of Mixed Modal Strategy

Symbol	Reading
$\exists^@x$	there exists an x such that ...
$\exists^\epsilon x$	there would then have existed an x such that ...
$F^{@@}xy$	x is thus-and-so relative to how y is
$G^{@\epsilon}xy$	x is thus-and-so relative to how y would then have been
$H^{\epsilon\epsilon}xy$	x then would have been thus-and-so relative to how y then would have been

There will also be plural quantifiers $\exists\exists^{\epsilon\epsilon}XX$. The relation of the modal logic involved to standard options in the existing literature is essentially the same for the present strategy as for the strategy of the preceding chapter.

PART III

Further Strategies and a Provisional Assessment

A

Miscellaneous Strategies

o. OVERVIEW

Chapter-length outlines of three specimen strategies of nominalistic reconstruction have now been given, the geometric strategy of Chapter II.A, a first (and purely) modal strategy in Chapter II.B, and a second (and mixed) modal strategy in Chapter II.C. Much briefer note will be taken of some other options in the present chapter. There are several variant or mutant versions of each of the three strategies already considered, and there are also intermediates or hybrids between some of them. A few specimens from the resulting zoo of strategies will be exhibited in section 1.

Nominalism may be contrasted with the other great heresy in mathematics, **constructivism**. While nominalism rejects all existence assertions in mathematics, constructivistic mathematics, in the broadest sense, rejects existence proofs of a certain kind, so-called 'non-constructive' existence proofs, which purport to establish that there exists a mathematical entity with some mathematical property, but do not even implicitly identify any specific instance of such an entity. (The traditional taunt of constructivistic mathematicians when presented with such proofs has been, 'That's not mathematics; it's theology!')

Constructivistic mathematics in the broadest sense is a loose grouping of schools. The constructive school in the narrower sense requires, before the existence of a real number will be admitted, that a purely mathematical specification of the number, such as will in principle permit the effective generation of the successive digits of its decimal expansion, must be given. This requirement is liberalized in one direction by **intuitionism**, which still requires a specification permitting the effective generation of successive digits, but does not require it to be purely mathematical, in that it allows real numbers generated by successive free choices of the creative mathematical subject. The requirement is liberalized in another direction by **predicativism**, which does not require a specification permitting the effective generation of successive digits, but does require a purely mathematical definition, and one avoiding certain vicious circularities.

Any close examination of the differences among different schools would be beyond the scope of this book, as would be any detailed discussion of the issue of the ultimate motivation for constructivism, which has often been represented as deriving, like the motivation for nominalism, from ontological considerations (and more specifically as deriving, unlike the motivation for nominalism, from some conceptualistic or idealistic or mentalistic or psychologistic view of mathematics). But one thing may be noted: whether or not it ultimately derives from ontological considerations, the motivation for constructivism is such that constructivists still continue to reject non-constructive proofs even when these are reconstrued along the lines of the strategies outlined in this book, so as to be proofs purely about concreta.

Thus less of standard mathematics is acceptable as it stands to nominalists than to constructivists, but more of it can be more easily made acceptable by reconstrual. When the negative, destructive sides of the two 'isms, their critiques of standard mathematics, are compared, constructivism appears more moderate than nominalism; but when their positive, reconstructive sides, their alternatives to standard mathematics, are compared, it appears more radical. Generally, contemporary proponents of constructivism have been opponents of nominalism, and vice versa. However, one point of contact between the two rival 'isms does emerge in a strategy sketched in the optional semi-technical appendix, article 1.f below.

It is characteristic of constructivist schools to wish to restrict standard logic, and of nominalistic strategies to wish to extend it. Strategies involving modal logic, plethynticology, and mereology have been considered already or will be considered in section 1. There are also some other strategies involving other extensions of classical logic, which will be discussed in brief in section 2. They will be discussed only briefly because the logics they involve are more problematic than those involved in the strategies considered so far. The logics deployed in the strategies considered so far may be called **domestic** logics. They are logics of notions present in ordinary language, though no formal counterpart of them was incorporated in the artificial language of standard logic (perhaps largely because standard logic was developed mainly for purposes of analysing mathematical proof, and the notions in question are not important in standard mathematics). Though these logics are not uncontroversial, they are logics for which it can with some plausibility be claimed that one already has an intuitive understanding of their basic notions.

The logics deployed in the strategies considered in section 2 may be called **imported** logics. They are logics of notions first defined in the specialist technical literature of mathematical logic. But nominalists reject orthodox mathematics in general, and must reject the orthodox definitions of these logical notions in particular, since under the orthodox definitions their deployment brings with it implications to the effect that certain abstracta exist. (A debatable example of this kind is metalogical modal logic, as discussed in article II.B.3.b.) The deployment by nominalists of such logics raises urgently the fundamental problem of what sanctions the acceptance of a logic, and the even more fundamental question of what constitutes the acceptance of a logic. These fundamental questions of philosophy of logic have not as yet been discussed at any length in the literature on nominalism (as they have been in the literature on constructivism), and not being discussed at much length in the literature that it is the primary aim of this book to survey, they will not be discussed at any length here.

But even without entering into such questions, it can be seen that there is something dubious about the practice of just helping oneself to whatever logical apparatus one pleases for purposes of nominalistic reconstruction, while ignoring any customary definitions that would make the apparatus nominalistically unpalatable: for by doing so, one can make the task of nominalistic reconstruction absolutely trivial—and so absolutely uninteresting. This is what will be suggested by the examples in section 2.

1. TAME STRATEGIES

a. *Between the Geometric and a Modal Strategy*

Any geometric theory that supposes space to be infinite or infinitely divisible supplies an infinity of disjoint entities, points. The second modal strategy could be followed down through the last step prior to the introduction of modality (at the beginning of article II.C.3.b), but then appeal to modality avoided in favour of an appeal to geometry. This would avoid the dependence on modality found in the strategy of Chapter II.C, and avoid the dependence on the specific details of a particular geometry found in the strategy of Chapter II.A. (Something similar could be done in the way of combining features of the strategies of Chapter II.B and Chapter II.A, given the relation between the strategies of Chapters II.C and II.B explained in article II.C.3.b.) Any geometric theory that supposes space to be continuous supplies uncountably- and continuum-many points, and so would permit the nominalistic reconstruction of even more of standard mathematics than just standard analysis, making use of the ideas of Lewis (mentioned at the end of section II.C.2).

One could go even further with these same ideas if one took space to be something more than continuous, so that ratios of distances between points would be appropriately measured not by real numbers but by, say, the **non-standard real** numbers of Abraham Robinson as in Keisler (1976) or the **surreal numbers** of John Conway as in Knuth (1974). (The strategy suggested here is one of the few in this book with no counterpart in the literature, and points to unutilized resources for nominalists.)

b. Between the Two Modal Strategies

A strategy so to speak between the two modal strategies is conceivable. The sequence of reductions of analysis from the second modal strategy could be stopped at some fairly early stage, where the most important apparatus would be natural numbers and sets thereof, with the natural order relation on natural numbers, and the elementhood relation between natural numbers and sets thereof. The leading idea of the first modal strategy, that of replacing assumptions of the existence of mathematical entities by assumptions of the possible existence of linguistic tokens denoting them, can then be brought in. The linguistic entities in this case would be finite numerals, denoting natural numbers, and infinite lists of such numerals, denoting sets of natural numbers. If a numeral for a natural number is taken to be a sequence of that number of strokes or ones, and a list of such numerals as a sequence of them separated by commas or zeros—or rather, if numerals and lists are taken to be physical entities somehow physically coding such sequences—then the objects that are taken as tokens on this strategy, and whose possible existence is assumed, would be the same as in the first modal strategy. The dynatontological commitments would be unchanged. But the ideological commitments would be lightened. One would have primitives only for relations like the counterpart for finite numerals of the order relation on numbers, essentially just 'is shorter than', and the counterpart of the elementhood relation between numbers and sets, essentially just 'occurs in'. Both 'shorter' and 'occurrence' are closely akin to 'part'.

c. Beyond the Second Modal Strategy

The second modal strategy, as compared with the first, has the benefit of lighter modal logical commitments, at the cost of heavier syndynatontological commitments, as explained in article II.C.3.b. A strategy beyond the second modal strategy in the direction of lighter logic and heavier syndynatontology is conceivable. The assumption of the second modal strategy was one of potential existence:

> There could have existed an infinity of disjuncta

One could go further and assume what may be called **perpotential** existence:

> There could have existed an infinity of disjuncta in addition to everything that actually does exist

One could go even further and assume what may be called **pluperpotential** existence:

> There could have existed an infinity of disjuncta in addition to everything that actually does exist, without anything that actually does exist being other than it is in any relevant respect

(despite the difficulties about this notion mentioned in article II.B.3.b). Of course, if there had been other things that there aren't, the things that there are would have been different in one respect at least: they would then have coexisted with those other things, as they now don't. The proviso about 'relevant respects' is intended to exclude this kind of respect.

Taking $\Diamond$ to read:

> it could have been the case, without anything there is ceasing to be or becoming other than as it is in any relevant respect, that . . .

would permit considerable simplification of the modal logic. Reverting to the notation of article II.C.3.b, the axioms of the theory would become:

(vi*) $\Diamond \exists \exists WW(WW$ are infinitely many $\land$ WW are nested)

(v*) $\Box \forall \forall WW(WW$ are nested $\rightarrow Q(WW))$

without the need for any changes in Q. For any theorem D that followed from the version of the theory as it was just before the introduction of modality, $\Diamond D$ will follow from the version of the theory as it is afterwards.

In order to deduce information about how things there are are, the appropriate modal logic would have to include a law:

$$\forall x \forall y \ldots (P(x, y, \ldots) \leftrightarrow \Diamond P(x, y, \ldots))$$

for all P that involve only quantifications $\exists x$, $\exists y$, . . . over things there are and only predicates Fxy about how things there are are—in other words, for all primary P. It is this law that distinguishes $\Diamond$ in the sense of plu-perpotentiality from $\Diamond$ in the sense of mere potentiality.

This law would be the only law involving a connection between how things are (on the left-hand side) and how they would have been (on the right-hand side), and the only law involving quantifying into a modal context. There is so little such 'quantifying in' that it might be an altern-ative to read $\Diamond$ as metalogical consistency, not in an absolute sense of mere logical consistency, but in a relative sense of consistency with the facts about how the things there are are (despite the difficulties about this notion mentioned in article II.B.3.b).

There may be a technical problem arising from the fact (mentioned in article I.B.2.b) that much of what science says about concrete entities is 'abstraction-laden'. It may be that not all information about how things are is expressed by primary formulas, and that one would have to intro-duce some new primary primitives, as on the geometric strategy. But one would not have to seek 'elegant' new primary axioms, since one would not, as on the geometric strategy, be attempting to reformulate the theory solely in terms of primary primitives and primary quantifiers.

There may be a philosophical problem in that the theory produced by this strategy may look a little too close to the bare assertion that concreta behave 'as if' abstracta existed and the standard theories about them were true. Whether this is indeed a problem depends on just what the grounds for dissatisfaction with the 'as if' theory are supposed to be.

d. Beyond the First Modal Strategy

There is an idea that seems to offer a way of going beyond the first modal strategy in the direction of lighter dynatontology and heavier ideology. The idea would be to try to reconstrue the version of analysis with nat-ural numbers and sets thereof, as in article 1.b, into a version with natural numbers and open formulas $F(\upsilon)$ about them, reducing a set $\{\upsilon \mid F(\upsilon)\}$ to an open formula $F(\upsilon)$ defining it. Then the leading idea of the first modal strategy would be brought in, and assertions about the existence of formula-types reconstrued into assertions about the possible existence of formula-tokens. Thus infinite linguistic tokens, lists of names of elements of a set, would be avoided in favour of finite linguistic tokens. In place of the notion of occurring on a list would come the notion of satisfying a formula. Once the idea of reducing sets or classes to open formulas or open sentences and elementhood to satisfaction is being contemplated, there seems to be little reason why it should be brought in only in the context of analysis and sets of natural numbers. Why not try, quite generally, to reconstrue sets of individuals in terms of formulas about individuals, and sets of sets in terms of formulas about formulas?

Now the alethic notions of truth and satisfaction are subject to certain difficulties: the famous **Epimenides** or **liar** paradox about the sentence asserting its own untruth, and its analogue, the famous **Grelling** or **heterological** paradox about the open sentence '*x* is an open sentence not true of itself'. The latter is the analogue of the famous Russell paradox about the set of all sets that are not elements of themselves. But that paradox is avoided by the standard and other hierarchical theories of sets, so the strategy would be to reconstrue one of these hierarchical theories along the lines indicated: the most obvious but not the only candidate for such treatment would be the end product of the logicist tradition, the stratified hierarchy of sets over infinitely many individuals (mentioned in article I.B.1.b).

The following sorts of entities are to be distinguished:

> the class or set of all red things
> the attribute or property of being red
> the open formula or sentence type '*x* is red'
> an open formula or sentence token '*x* is red'

There is considerable precedent within the logicist tradition for conceiving of sets as somehow derivative from something so to speak intermediate between sets and formula-tokens. The obstacle to reconstruing assertions of the existence of sets into assertions of the existence of formula-tokens is that there do not in fact exist infinitely many formula-tokens. But reduction aside, there is also an obstacle to the reduction of standard mathematics to the stratified hierarchy of types over infinitely many individuals in the simple fact that there may well not exist infinitely many concrete individuals. There is considerable precedent within the logicist tradition for surmounting this latter obstacle by weakening the assumption of the actual existence of an infinity of individuals to the assumption of the possible existence of an infinity of individuals, and reconstruing categorical assertions about how things are into hypothetical assertions about how things necessarily would have been if there had existed an infinity of individuals. It would involve only a short step beyond the logicist tradition to attempt to surmount the former obstacle in the same way. The result would be just the kind of strategy of nominalistic reconstrual of sets and elementhood in terms of possible tokens and satisfaction contemplated above. It is in this sense that the logicist version of a hierarchical theory of sets more than any other invites treatment by this strategy.

The ideological commitments of such a strategy must now be assessed, especially since many nominalists profess **physicalism** in a sense involving both ontological and ideological restrictions. Ontologically, only physical entities are accepted. Ideologically, only physical predicates are accepted. Of course, whether a nominalist ought to be a physicalist of any variety, and so ought to have serious doubts about the strategy being contemplated, depends on just what the motivation for sympathy with nominalism is supposed to be. Now while formula-tokens may be physical entities, satisfaction is not directly a physical relation with those entities, and truth is not directly a physical property of them. Rather, these are linguistic, and more specifically alethic, properties and relations. There is serious doubt that these notions can be defined or explained purely in terms of physical notions. (See in this connection Field (1972, 1986).)

To be sure, ideological restrictions on acceptable predicates include both restrictions on what kinds of predicates are acceptable as undefined primitives, and restrictions on what kinds of logical operators are acceptable for defining further predicates from given ones; and many versions of 'physicalism' conceive of it as a restriction on primitive predicates, but not on logical operators. So if alethic notions could be argued to be quasi-logical operators, they might be physicalistically acceptable, at least for many versions of physicalism. On a disquotational theory of truth, there is some hope of arguing that the truth predicate is, like the device of quotation it undoes, a logical or quasi-logical operator. But (as mentioned in article I.A.2.d) a disquotational theory of alethic notions is a local one, applying only to a single language, one's own; and the strategy being contemplated requires a global notion of satisfaction, one applicable to all possible languages, or at least to an indefinitely wide range of them.

e. Beyond Both Modal Strategies

A combination of the leading ideas of the strategies of articles 1.c and 1.d—satisfaction and pluperpotentiality—is also conceivable. Producing a token of an open formula that is satisfied by some entities and not others is just one way of mentally distinguishing the former from the latter, or collecting the former in the mind. And if there are any sort of entities for which pluperpotentiality is an at all plausible assumption, for which it is an at all plausible assumption that they could have existed without any physical entity that actually exists being in any physical respect other than as it actually is, the most obvious examples of such entities would be minds. These observations suggest a reconstrual of any theory of a hierarchy of sets in terms of a hierarchy of minds (presumably disembodied and not subject to human limitations). The lowest-level minds would mentally collect physical individuals. Then higher-level meta-minds would mentally collect lower-level minds. Or instead of multiple minds, one could consider mental acts of collection on the part of some one supreme Mind. The result would be a reconstrual of mathematical theorems as assertions of the form:

> Mind could have existed, and
> if Mind had existed then It would have . . .

With the partial exception of the strategy of the article 1.f to follow, apparatus unacceptable to constructivists is present in all the strategies considered so far. However, the chasm separating constructivism and nominalism is perhaps most conspicuous in the strategy we have just been contemplating. That strategy especially invites the traditional constructivist taunt, 'That's not mathematics; it's theology!' For what are these levels of minds and meta-minds, and Mind, but choirs of angels and archangels, and God?

f. Beyond the First Modal Strategy, bis

The alternative to the strategy of article 1.d that tries to make do with only local alethic notions, applying to only one language, invites exploration, in which case it will be convenient to return to the context of analysis, and consider how it might be initially reconstrued in terms of formula-types, themselves intended later to be reconstrued in terms of possible formula-tokens. Ideas from intermediate-level logic can be of use in various ways in connection with the kind of strategy being contemplated. One such idea is the coding of formulas by natural numbers (alluded to in article I.B.4.b). Other such ideas enable one to define the two-place satisfaction predicate in terms of syntactic substitutions and a one-place truth-predicate:

> the open formula $F(x)$ is true of or satisfied by the natural number
> n if and only if
> the closed formula $F(\underline{n})$ is true, where $F(\underline{n})$ is

the result of substituting the numeral $\underline{n}$ for the number n in place
of the free variable x

Suppressing technical details, the result is that the strategy being contemplated now takes the following form: start with a theory T_0 in a language L_0 with variables for natural numbers and primitives for the usual arithmetic relations on them, with the usual arithmetic axioms. Add predicate V_0 for:

ξ is (the code number of) a formula of L_0 that is true

with reasonable axioms, to obtain a theory T_1 in a language L_1. Then look how much of the usual theory of sets of natural numbers, with axioms of extensionality, comprehension, and choice, can be reduced to this theory and language. Now even without entering into technical details, it is probably intuitively fairly clear that the theory of all arbitrary sets of natural numbers cannot be reduced to the theory and language being contemplated. When technical details are entered into, it turns out that what can be so reduced is a restricted theory of sets with axioms of extensionality and restricted comprehension:

$$\exists X \forall \xi (\xi \in X \leftrightarrow P(\xi)) \text{ for } P \text{ not involving set variables}$$

One could, of course, go on to add a truth predicate V_1 for formulas of L_1, to obtain a theory T_2 in a language L_2, getting some more sets. One could, indeed, go on for quite some time this way, getting more and more sets. In the end, what one arrives at in this way is not classical analysis but rather what is known as **predicative analysis** (or in the more general context of hierarchical set theory, at what is known as the **ramified theory of types**, the general framework for predicativist mathematics).

There may be some doubts as to whether predicativism supplies enough mathematics for applications to sophisticated physics. However, its scope and limits have been intensively investigated for several decades, especially by the school of Solomon Feferman, and as part of a more general study of related 'isms by the school of Harvey Friedman; and the work of these schools provides some grounds for optimism. There has also been much work in the last fifteen years or so on languages containing their own truth predicates, the work of Saul Kripke being particularly influential. More recently connections have been established between this work on truth and the work on predicativism. All this technical work represents a heretofore unexploited resource for nominalists. (The survey Sheard (1994) would be a good place for the interested reader to start.)

2. WILD STRATEGIES

a. Substitutional Strategies

Leśniewski, originator of mereology, was also the originator of another and more controversial logical device, called **substitutional quantification**. It is in fact not a single device but a whole family of them, one for each grammatical category. The grammatical category most pertinent for present purposes is that of open formulas with one free variable. For that grammatical category, the 'syntax' of substitution operators is fairly easily explained. Let $L°$ be a one-sorted language with variables x, y, ... for basic entities. Let L^ε be the two-sorted theory in the two-sorted extension L^ε of $L°$ with variables X, Y, ..., with the primitive $\in$. The **substitutional** language L^Σ is just like in L^ε except that $x \in X$ is written $X(x)$ and $\exists X$, $\exists Y$, ... are written ΣX, ΣY, ... The symbol 'Σ' is called a **substitutional quantifier**. The axioms of substitutional logic are the rewritten versions of extensionality and restricted comprehension as in article 1.f. The **semi-substitutional** language L^σ is just like L^ε except that $x \in X$ is written $X(x)$ and $\exists X$, $\exists Y$, ... are written σX, σY, ... The symbol 'σ' is called a **semi-substitutional quantifier**. The axioms of semi-substitutional logic are the rewritten versions of extensionality, comprehension, and choice as in article II.C.1.a. Higher and higher levels of substitutional and semi-substitutional quantifiers may be introduced, producing rewritten versions of the ramified and the simple theories of types, and of predicativist and classical mathematics.

Would thus rewriting mathematics in substitutional terms constitute a nominalistic reconstrual of it? If so, then one has a very easy route to such a reconstrual. Whether it is so depends on the 'semantics' of substitution operators. What is the meaning of the operators Σ, σ? If a way of reading them is wanted, 'there substists' and 'there semi-substists' suggest themselves. But a way of pronouncing is not in itself a way of understanding. What is the nature of the commitments these operators involve? If a solemn term is wanted, these may be called **hypocatastatic** and **hemi-hypocatastatic** commitments. But a label is not an explanation. The orthodox doctrine on substitutional quantification is authoritatively stated in a paper of Kripke (1976). Very roughly, the orthodox definition would read:

(i) $\Sigma X(\ldots X \ldots)$ is true if and only if
            there exists a formula $Q(x)$ of L
            such that $\ldots Q(x) \ldots$ is true

Kripke does not treat semi-substitutional quantification, but the analogous definition would read:

(ii) $\sigma X(\ldots X \ldots)$ is true if and only if
     there exists a formula $Q(x)$ of some extension L' of L
     such that $\ldots Q(x) \ldots$ is true

With these definitions, the substitutional and semi-substitutional axioms can be proved true in orthodox mathematics. But plainly with these explanations and understandings, use of the substitutional and semi-substitutional operators, though it does not involve the assumption of the existence of sets, does involve the assumption of the existence of infinitely many formulas, presumably abstract expression types, and ideological commitment to notions of satisfaction or truth, local in the one case, global in the other.

Substitutional operators are sometimes described as devices of 'infinitary disjunction'. Now they are not themselves infinitary notations, else it would have been impossible for us to write them down above. They may be introduced as finite *abbreviations* for infinitary disjunctions, provided the language into which they are being introduced already includes infinitary disjunctions for them to abbreviate. But no language spoken or written by human beings does. Human beings can indeed in some cases get an understanding of how certain operators would function in a language that it is a 'medical impossibility' for human beings to speak or write. In particular, they can get an understanding of how infinitary disjunctions would function from definitions comparable to (i) above: the disjunction of infinitely many sentences would be true if there was some sentence that was a disjunct of it and was true. But gestures in the direction of infinitary disjunction do nothing to show how reliance on definitions mentioning expression types, truth, and so on, can be avoided.

Alternative definitions would avoid the assumption of the existence of formula-types in favour of the assumption of the possible existence of formula-tokens, roughly:

(i*) $\Sigma X(\ldots X \ldots)$ is true if and only if
     there could have existed a formula $Q(x)$ of L
     such that $\ldots Q(x) \ldots$ was true

and analogously:

(ii*) $\sigma X(\ldots X \ldots)$ is true if and only if
     there could have existed a formula $Q(x)$ of some extension L' of L
     such that $\ldots Q(x) \ldots$ was true

Then plainly, with these explanations and understandings, the sub-stitutional or semi-substitutional rewriting of predicativist or classical mathematics would amount to the strategy of articles 1.d and 1.f in a different notation. Substitutional and semi-substitutional logic would not be providing an alternative to modal logic in a project of nominalistic reconstruction, but just an abbreviation for it.

There remains the alternative of taking substitution operators to be primitives not requiring explanation in terms of anything more familiar, and their laws to be axioms not requiring justification in terms of any-thing more fundamental. This is the kind of alternative that was said in section 0 to raise problems. It is, however, the position of some advocates of substitutional logic. For some advocates of substitutional logic advance the slogan that 'substitutional quantifiers must be used in the meta-language in stating the semantics of substitutional quantification'. What this slogan amounts to is that while (i) and (ii) are rejected, something like the fol-lowing are accepted:

(i⁻) $\Sigma X(\ldots X \ldots)$ is true if and only if
        there substists a formula $Q(x)$ of L
        such that $\ldots Q(x) \ldots$ is true

and analogously:

(ii⁻) $\sigma X(\ldots X \ldots)$ is true if and only if
        there semi-substists a formula $Q(x)$ of some extension L' of L
        such that $\ldots Q(x) \ldots$ is true

What the slogan amounts to is an obscure way of expressing the refusal to explain substitution operators in any more familiar terms. What the position advocated in the slogan amounts to is a paradigm of problematic primitivism.

b. Functorial Strategies

Leśniewski, originator of mereology and substitutional quantification, was also one of the precursors of even more controversial logical devices, called **predicate functors**. A little background is required before they can be introduced.

Nominalists are very moderate compared to **monists**, whose view is summarized by Prior on the last page of his major work (Prior 1969) on tense logic: 'there is only a single genuine individual (the Universe) which gets John-Smithish or Mary-Brownish in such-and-such regions for such-and-such periods'. (David Lewis has pointed out that Prior really should say, '. . . regionally and periodically'.) And Mary Brown and John Smith don't do anything: the Universe does things Mary-Brownishly and John-Smithishly. Monists in turn are moderate in comparison to **nihilists**, who deny the existence of anything at all. But if it is permitted to help oneself to whatever logical apparatus one wants, while ignoring the usual definitions and giving no other explanations, then not only a nominalistic but a monistic and even a nihilistic reconstrual can very easily be given, not only for mathematics, but for anything at all. The logical apparatus required for this reconstrual has been provided by—of all people— Quine, building on the work of Leśniewski and other precursors. For the author of the slogan that 'to be is to be the value of a variable' also tells us in one of his papers (Quine 1960*b*) how 'variables can be explained away'.

Quine proceeds, first, to extend standard logic by the addition of new operators called predicate functors, introduced as follows.

Predicate Functors

Symbol	Definition
$(\nu F)x_1 \ldots x_m$	$\sim Fx_1 \ldots x_m$
$(\kappa FG)x_1 \ldots x_m y_1 \ldots y_n$	$Fx_1 \ldots x_m \wedge Gy_1 \ldots y_n$
$(\varsigma F)x_1 \ldots x_{m-1}$	$\exists x_m(Fx_1 \ldots x_{m-1}x_m)$
$(\rho F)x_1 \ldots x_{m-1}$	$Fx_1 \ldots x_{m-1}x_{m-1}$
$(\phi F)x_1 \ldots x_{m-1}x_m$	$Fx_1 \ldots x_m x_{m-1}$
$(\psi F)x_1 \ldots x_{m-1}x_m$	$Fx_m \ldots x_{m-1}x_1$

Each corresponds to a certain modification of a verb, and so to an adverb in a generalized sense. This is shown by the examples in the table below, which suggest a way of pronouncing predicate functors.

Readings of Predicate Functors

Symbol	Definition
$(\nu$ talks$)x$	x *doesn't* talk
$(\kappa$ walks runs$)xy$	x and y *respectively* walk and run
$(\varsigma$ stares at$)x$	x (*just*) stares
$(\rho$ destroys$)x$	x *self*-destructs
$(\phi$ eats$)xy$ or $(\psi$ eats$)xy$	x *suffers* or *undergoes* eating by y

Quine proceeds, second, to show how to eliminate variables and quantifiers in favour of these new functors. In this way, starting with an English original, regimenting, symbolizing, eliminating other classical symbols in favour of ~ and ∧ and ∃, eliminating all classical symbols in favour of functors, desymbolizing, and pronouncing as suggested above, one obtains a quasi-English reconstrual:

> whatever lives, changes
> $\forall x(x$ lives $\rightarrow x$ changes)
> $\sim\exists x(x$ lives $\wedge \sim(x$ changes$))$
> $\sim\exists x(Fx \wedge \sim Gx)$
> $\sim\exists x(Fx \wedge (\vee G)x)$
> $\sim\exists x(\kappa F(\vee G))xx$
> $\sim\exists x(\rho(\kappa F(\vee G)))x$
> $\sim(\varsigma(\rho(\kappa F(\vee G))))$
> $\vee(\varsigma(\rho(\kappa F(\vee G))))$
> $\vee(\varsigma(\rho(\kappa$ lives $(\vee$ changes$))))$
> $\vee(\varsigma(\rho(\kappa$ lives (doesn't change)$)))$
> $\vee(\varsigma(\rho($respectively live and don't change$)))$
> $\vee(\varsigma($self-respectively lives and doesn't change$))$
> $\vee($just self-respectively lives and doesn't change$)$
> doesn't just self-respectively live and not change

Quine does not provide rules for carrying out deductions directly in the notation of predicate functor logic, but such have been supplied rather elegantly by Bacon (1985).

Monists will wish to supply a subject, the Universe, for the foregoing quasi-English reconstrual. Nihilists won't. Nominalists have the option, given a two-sorted theory with one style of variables for concreta and another for abstracta, of applying the kind of transformation indicated only to variables of the second sort, keeping those for concreta and eliminating those for abstracta. Would thus rewriting classical mathematics in functorial terms constitute a nominalistic reconstrual of it?

If so, then one has an almost trivial route to a nominalistic reconstrual. Needless to say, a nominalist could not just deploy functorial logic with *Quine's* translations into standard logic as explanations of what the functors are supposed to mean since, as Quine emphasizes, those explanations in effect attribute to formulas of functorial logic the same existential import as the formulas of standard logic they replace. What about deploying functorial logic without any translation into any more familiar logic? To do so would be another paradigm of problematic primitivism. If functorial primitivism is a less popular kind of primitivism than substitutional primitivism, it is perhaps because it is a more conspicuously problematic kind.

c. Diacritical Strategies

There is an even more controversial logic sometimes met with in the literature that may be called **diacritical** logic, from the Greek for 'to distinguish'. What this distinguishes is a narrower realm of what there exists, with an existential quantifier 'there exists an x' or $\exists x$, and a broader realm of what there is, and a being quantifier 'there is an x' or $\$x$. This distinction enables the diacritic, when talking to nominalists, to sound hard-headed by saying, 'Of course there exist no numbers'; and when talking to anti-nominalists to sound broad-minded by saying, 'Of course there are numbers'; and when talking to both at once to propose a compromise.

Actually, diacritique comes in two versions: the less spectacular ascribes to $\$$ the very same deductive rules applying to $\exists$. The more spectacular has to modify these deductive rules, because in addition to ordinary mathematical entities it allows for the being (though not the existence) of **impossibilia**: in addition to the ordinary square square of mathematics, it acknowledges an **inconsistent object**, the extraordinary round square; and in addition to the ordinary triangles of mathematics, each of which is either equilateral or isosceles or scalene, it recognizes also an **incomplete object**, the extraordinary general triangle, which is none of the three. The result is that the spectacular version of diacritical logic has to be in addition a **dialectical** logic, in the sense of a logic that tolerates—indeed, celebrates—contradictions.

Neither version of diacritique has played any very large role in mainstream philosophy of mathematics. Anti-nominalists are likely to find the spectacular version of diacritique unacceptable; and they may find even the less spectacular version dubious, since it still rejects orthodox mathematics, inasmuch and in so far as that subject is full of existence and uniqueness theorems (not being and uniqueness theorems). Nominalists are likely to reject both versions as well, since their attitude towards abstracta is that they 'reject them altogether'; it is not that they reject them as purported existents but accept them as putative beings: the commitments to abstracta nominalists wish to avoid are ontological (from the Greek for 'to be') and not just **hyparxological** (from the Greek for 'to exist'). But the most fundamental objection, on the part of both nominalists and anti-nominalists alike, is that the distinction it proposes is a distinction without a difference.

Simply rewriting ∃ as $ is easy, but understanding what the difference is supposed to be is not. Nor does it seem helpful or clarifying to speak not of a 'being' quantifier 'there is an x' but of a 'particular' quantifier 'for some x'; or to bring in 'subsistence' and 'reality'; or to switch from distinguishing 'is' and 'exists' to distinguishing 'thing' and 'entity'; or to bring in 'item' and 'object'; or to distinguish not different terms but different senses of the same term; or finally—surely a move of desperation, this—to switch from ordinary English to philosophical German and tell of the *Aussersein* of *Gegenstände*. Even sympathizers with substitutionalism or functorialism are likely to find in diacriticism a case where it would be illegitimate to accept a logical operator without some further explanation: even those who can swallow Σ or ς are likely to stick at $. In a sense, diacritique constitutes a kind of *reductio ad absurdum* of the free acceptance of imported logics.

B
Strategies in the Literature

0. OVERVIEW

Almost every one of the specimen strategies presented so far has been adapted from a source in the literature, from an attempt by some actual nominalist to show how standard scientific theories can be reconstrued nominalistically. The brief survey of the relevant literature to be undertaken in this chapter is intended to serve both as an acknowledgement of sources for Part II and Chapter A of this part and as a guide for further reading. A geometric strategy is treated in section 1; modal strategies are treated in section 2; miscellaneous further strategies are treated in the optional appendix section 3. With a couple of exceptions, only projects that have led to books or accumulations of articles of comparable bulk are covered, and most technical details and philosophical subtleties connected with the many strategies mentioned are omitted. In the article headings, the strategy or strategies in this book most relevant to the author or authors being discussed are indicated. Works specially recommended for readers wishing to explore the original sources are starred, beginning with Goodman and Quine (1947)*.

1. A GEOMETRIC STRATEGY

a. Field [cf. Chapter II.A]

Geometric nominalism has had only one important advocate, Hartry Field, but his work has generated so much discussion that in assigning credit (or blame) for the contemporary prominence of the issue of nominalism, Field must be named immediately after Quine, Goodman, Benacerraf, and Putnam. His relevant works include two books. The first launched his project. The second, Field (1989), begins with an introductory survey of the issue as chapter 1, and then reprints a half-dozen papers from the 1980s directly or obliquely related to the project as chapters 2–7. Chapter 2, a reprinting of Field's position paper Field (1982)*, provides a convenient summary of the first book. There have been several further papers in varying degrees related to the project since the second book, possibly the beginnings of the accumulation of materials for a third book.

In his first book, Field shows how to produce a nominalistic or synthetic reconstruction for a standard or analytic theory from classical physics, the extension of the method to post-classical physics being left an open problem for future research.

Actually, he produces not one but two synthetic reconstructions for a given analytic theory from classical physics, which he calls the 'first-order' and 'second-order' options. The second-order option is not literally based on 'second-order' logic as this is usually understood in the literature, but rather on a nominalistic substitute. Mereology is taken as this substitute in the book, and plethynticology in later papers. (Compare section II.C.1.) In the book, Field avoided committing himself to one rather than the other of his two options as his official theory. Subsequently he has almost definitely committed himself to the first-order version, and so the second-order version will not be further considered here. The first-order option is not quite literally based on 'first-order' or standard logic alone, since there is an incidental use of finite cardinality comparison logic in his treatment of proportion along traditional lines. The use of this extension of standard logic can in fact be avoided, essentially just by substituting the modern for the traditional definition of proportionality (as discussed in article II.A.3.d).

In establishing the relationship between an orthodox theory and the nominalist substitute for it, Field's procedure is roughly as follows. He does not directly reconstrue or reinterpret the analytic theory T_1 in the synthetic theory T_0, but proceeds indirectly in attempting to establish that T_0 is indeed an adequate replacement for T_1. He takes the primary condition of adequacy to be that T_1 should be conservative over T_0 in the sense that anything expressible in the language of T_0 and provable in T_1 should be provable in T_0. To establish this, he introduces an intermediate theory $T_0^\dagger$ and then first reinterprets T_1 in $T_0^\dagger$, and second shows $T_0^\dagger$ is conservative over T_0. The detour through the intermediate theory can be avoided by making explicit a direct reinterpretation of T_1 in T_0 that is implicit in the work done at the first stage.

Field's own informal descriptions of his method, beginning with the opening chapter of his first book, generally down-play the dependency of his strategy for its success on the presumed fact or expert opinion that all the mathematics required for physical applications to date can be developed on a much more restrictive basis than that of the standard axioms of set theory. Rather, his formulations put heavy emphasis on what goes on at the second stage in his (avoidable) two-step approach, which he takes to illustrate a general phenomenon, roughly to the effect that:

the addition of mathematical axioms to a nominalistically-stated theory makes it easier in practice to draw nominalistically-stated conclusions, but the conclusions we arrive at by these means are not genuinely new, and could in principle have been arrived at working in the nominalistically-stated theory alone.

(This is not a direct quotation of a single passage from Field's work, but a pastiche of several.) Such formulations have proved quite controversial, but since the controversy involves technical issues it might be considered inappropriate to enter into it here. The technical situation is briefly described in the optional semi-technical appendix, article 1.b below.

A variant version of Field's approach, motivated in part by unpublished criticisms of Field's work by Kripke, was given in Burgess (1984*b*), since superseded by Burgess (1991)* (supplemented by Burgess (1990*a*), an indirect response to Field (1985*b*), itself reprinted as chapter 6 in Field's second book). This variant emphasizes the dependence of the strategy on the expert opinion that enough mathematics for applications is provided by analysis, gives a direct reinterpretation of the analytic theory in the synthetic, and uses the modern theory of proportionality and only standard logic. It was the immediate source of Chapter II.A of the present book, for which Field's work is therefore the source at one remove.

The indirectness of Field's approach raises a subtle issue about metalogic. Suppose van Cee is a physicist who professes to exclude abstracta from his ontology, while von Dee is a physicist who includes them in hers. Von Dee accepts an analytical physical theory T_1, while van Cee rejects T_1 and accepts only a synthetic physical theory T_0. Suppose also that bar Gimel is a metalogician who rejects abstracta, while ben Daleth is a metalogician who accepts them. Ben Daleth has developed a standard metalogical theory M_1, while bar Gimel has so far taken only the first steps towards developing a nominalistic metalogical theory M_0. Suppose now that von Dee produces a deduction S_1 from T_1 of some interesting result P_0 expressible in the language of T_0. And suppose ben Daleth deduces from M_1 a metatheorem to the effect that whenever there is a deduction from T_1 of a result expressible in the language of T_0, then there is a deduction of the same result from T_0. If this metatheorem is true, then if van Cee works long enough, he can eventually find such a deduction S_0 and so arrive at the result P_0. But surely van Cee cannot simply conclude on the strength of the existence of the deduction S_1 from T_1 that there must also exist a deduction S_0 from T_0, and in this way arrive at once at the conclusion P_0, without actually having to look for such a deduction. He cannot do this because he is not in a position to conclude that the metatheorem is true. He could only conclude that if bar Gimel's M_0 were developed to the point that it yielded the same metatheorem as ben Daleth's M_1.

(This issue should not be confused with another, related one. Suppose a philosopher Gammapoulos, who professes to exclude abstracta from his ontology, and another philosopher Deltaki, who includes them in hers, are debating about what van Cee can accomplish as contrasted with what von Dee can accomplish. And suppose Deltaki claims that van Cee will never be able to deduce the result P_0. Presumably, Gammapoulos could legitimately produce a *reductio* of Deltaki's position by pointing out that unlike himself, she professes to accept the standard metalogic M_1, and according to that standard metalogic, van Cee will eventually be able to deduce the result P_0.)

While Field has not explicitly said that the development of a nominalistic metalogic is required for the sake of nominalistic physics, he has said that the development of a nominalistic metalogic is required for its own sake. His later papers (Field 1991, 1992) deal with philosophical and technical issues connected with metalogic, elaborating on ideas from earlier papers (Field 1984a, 1988), reprinted as chapters 3 and 7 in his second book. (In these cases there are significant additions and/or amendments in the reprinting, so the second book supersedes the separate papers.) While Field considers it appropriate to invoke a geometric apparatus in a nominalistic treatment of the natural science of physics, the science of space and time, he considers it inappropriate in a treatment of the formal science of metalogic. The employment of different strategies for different sciences (physics, metalogic) is a unique feature of his approach.

The apparatus he does invoke in connection with metalogic includes two extensions of standard logic: a logic of metalogical modality, and a logic of substitutional quantification (or alternatively, of recursive infinitary conjunction and disjunction). He takes the relevant operators as primitives, rejecting the orthodox definitions of the relevant notions, since these involve abstracta, and rejecting alternative definitions in terms of metaphysical modality, since he rejects metaphysical modality also. His rejection of metaphysical modality is an almost unique feature of his approach.

So, too, is the degree of explicit attention he gives to the fundamental question of just what the acceptance of a logic consists in, which too many nominalists have neglected. But Field's work in this direction has not yet reached its final form, and his project remains ongoing.

b. Field's Critics

There is a large secondary literature on Field's work. A good example (in both senses) of philosophical criticism of the assumption of geometricalia (a strategy peculiar to Field) and of extended logics (a strategy common to all reconstructive nominalisms) is provided by Resnik (1985*a*, 1985*b*). As to less purely philosophical criticisms turning on technical issues, the first important published item was Shapiro (1983*b*), to which Field replied in Field (1985*a*), reprinted as chapter 4 in his second book. Alongside Field's own books may be mentioned the volume of conference proceedings (Irvine 1990) edited by Andrew Irvine; it would be only a slight distortion to call it an anthology of commentaries on Field. Alasdair Urquhart's contribution (Urquhart 1990*) to that volume is the most authoritative published source for technical matters. It takes account of Field's first book and subsequent work down to the mid-1980s. Penelope Maddy's contribution (Maddy 1990*b*) to the same volume also addresses some technical matters pertaining to the second-order version of Field's approach, which has not been discussed here, as does Maddy (1990*c*). It is in replying to such criticisms in Field (1990) that Field comes closest to abandoning the second-order version entirely and adopting the first-order version as the official one. The works cited have influenced the following account of the technical issues (and article II.A.5.b). (Also in the same volume are critical works by members of the St Andrews School, Hale (1990) and Wright (1990); these are part of a long exchange that began with Field (1984*b*), reprinted as chapter 5 in Field's second book, and continues to this day.)

The technical issues in part turn on a rather subtle distinction pertaining to axiom **schemes**. When a theory T in a language L involves a scheme, we have said that we conceive of a scheme as a **rule** to the effect that:

(i)   for every formula Q, $—Q—$ is an axiom

An alternative would be to conceive of it as the **list** of all the axioms:

(ii) $—Q_0—$, $—Q_1—$, $—Q_2—$, . . .

that result when this rule is applied to all formulas:

(iii) Q_0, Q_1, Q_2, . . .

of the language L. The difference in conceptions makes a difference only
when the theory T in the language L is extended to or incorporated in
some stronger theory T' in some richer language L' which will have new
formulas Q' not on the list (iii), and for each such formula a new formula
—Q'— not on the list (ii). To conceive of the scheme as a rule (i) means
that these new formulas —Q'— are taken as new axioms; to conceive of
the scheme as a list (ii) means that they are not.

Returning to the description of Field's strategy, his intermediate theory
may be described as follows. The synthetic theory T_0 involves a scheme
of continuity. The intermediate theory $T_0^\dagger$ Field considers is the result of
adding the apparatus of set theory while treating that scheme as a list (and
contrasts with $T_0^\ddagger$ considered in article II.A.5.b, the result of adding the
apparatus of set theory while treating the scheme as a rule). The first step
on his two-step approach is, as indicated in article 1.a, in effect to reinterpret T_1 in $T_0^\dagger$, replacing quantifications over real numbers in T_1 by
quantifications over sets of some kind in $T_0^\dagger$. More is provable from the
standard axioms of set theory about the real numbers as standardly reconstructed in set theory than is provable about real numbers in analysis. For
present purposes, it is only necessary to reinterpret quantifications over
real numbers in T_1 by quantification over sets of some kind for which the
axioms of analysis can be shown to hold. It is not necessary to reinterpret
quantifications over real numbers in T_1 by quantifications over the real
numbers as standardly reconstructed in set theory. Nor is it possible to do
so working in $T_0^\dagger$ (as it would be working in $T_0^\ddagger$). The reinterpretation
that works is one that replaces quantification over real numbers in T_1 by
quantification over equivalence classes of ordered triples of points under
the equivalence relation of proportionality. This is Field's approach.
Implicit in the proof, but not explicit in the statement, of the reinterpretation or representation theorem he thus proves is a direct, one-step
reinterpretation of T_1 in T_0, replacing quantifications over real numbers
by threefold quantification over points. As Urquhart says, 'if you look
inside the proof of the representation theorem to see what makes it tick',
what will be found is the direct reinterpretation.

The second step in Field's two-step approach is to show that $T_0^\dagger$
is conservative over T_0 (as $T_0^\ddagger$ is not, since adding standard set theory
gives new geometric results, and adding proposed further axioms beyond
standard set theory gives further geometric results, with the majority and

minority proposals as to further axioms leading to incompatible geometric results). In fact, adding the apparatus of set theory to any theory produces a conservative extension, provided any and all schemes in the original theory are treated as lists, not rules. This is one of a cluster of results reported in an appendix in Field's book (the most interesting perhaps being the one there attributed to Scott Weinstein). Field sums up the situation in controversial semi-colloquial formulations like the one indicated in article 1.a. We hope and believe that enough relevant information has been given in this section (and article II.A.5.b) to allow readers to form a judgement about whether this semi-colloquial formulation is insightful or misleading, or at any rate, to free readers from having to depend on semi-colloquial formulations for their understanding of the matter.

2. MODAL STRATEGIES IN THE LITERATURE

a. *Chihara I [cf. article A.1.f]*

One of the earliest large-scale nominalist projects was that undertaken in Charles Chihara's first book (Chihara 1973). The book begins with a striking testimonial to the importance of Quine's work as an inspiration, or provocation, to later nominalists. Its introduction opens by quoting an anti-nominalist passage from one of Quine's works, and continues: 'When I first read these words in [Quine's] *Word and Object* several years ago, I wrote in the margin: "This philosophical doctrine should be soundly refuted." ' There is a polemical chapter against Quine (containing a survey of the first quarter-century of debate over Quine's views on 'ontological commitment') and another against Gödel. There are historical chapters on Russell's logicism and Poincaré's predicativism (which partially document some of our passing historical remarks in articles A.1.d and A.1.f).

From these there emerges in the final chapter the outline of a nominalistic reconstrual of at least a significant part of predicativist mathematics, Chihara (1973: chapter 5)*, with a sketch (but only a sketch) of how to reconstrue mixed mathematico-physical language. A more rigorous, technical presentation is given in an appendix. Article A.1.f is essentially a sketch (but only a sketch) of Chihara's approach. Chihara has not sought to determine what are the outer limits of predicativism, or just how much of applicable mathematics can be developed within them. Any attempt at such a determination at the time he was writing his first book would soon have been overtaken by the rapid progress being made in the studies alluded to in article A.1.f. Chihara has not attempted such a determination in more recent studies, because after the publication of his first book he became convinced that the restrictions of predicativism are unnecessary. He thereby became, for purposes of this survey, another philosopher.

b. Chihara II [cf. article III.A.1.d, Chapter II.B]

Chihara's new position was outlined in a longish paper, Chihara (1984)*, and detailed in the first half of his second book, Chihara (1990). Article A.1.d is essentially a sketch of Chihara's new approach. In making comparisons, it should be noted that the terminology used in Chihara's second book differs from that used in his first book and in this one. He drops the label 'nominalist', apparently because some have understood nominalism as consisting not in the ontological rejection of abstracta, but rather of this plus extensionalist rejection of modality and other ideological aspects. He even tends to avoid the label 'modal', lest anyone make the—unintelligent—assumption that anyone who speaks of 'modality' is committed to existing systems in the standard literature. His new label for his own position is 'constructibilism', which may itself be liable to confusion with 'constructivism', and so it should be noted that his new, 'constructibilist' position has no connection with constructivism (whereas his old, predicativist position did have such a connection).

The exposition of the new strategy in the second book is much more detailed than that of the old strategy in the first. The reduction of sets of sets to formulas about formulas involves subtleties not present in the reduction of sets of individuals to formulas about individuals, and these are given due attention. Above all, the exact nature of modal logic involved, which is crucial to the reconstrual of mixed, mathematico-physical language, is worked out in some detail (with acknowledgements to earlier work of Ernest Adams). The same modal logic is detachable from the strategy of reinterpreting type theory in terms of formulas, and usable in other nominalistic strategies. It is in this sense that Chapter II.B of this book is inspired by Chihara's work.

The presentation of that modal logic in Chapter II.B has by design been complementary to that in Chihara's book. Chihara provides an axiomatization of the logic on its own terms, while Chapter II.B attempted to relate that logic to existing systems in the standard literature. Chapter II.B attempted an informal explanation of the logic avoiding mention of 'possible worlds', while Chihara allows himself to use the language of 'possible worlds', though as a nominalist he is no believer in such alleged entities, because many philosophers find it aids in understanding. Now quite a few philosophers have wished to speak of 'possible worlds' while denying they seriously believe in them. Such philosophers generally adopt one or the other of two options, called **ersatzism** or **fictionalism**: the former reconstrues 'possible worlds' in terms of abstract representations of how the actual world might have been. The latter distances itself from talk of 'possible worlds', declaring it to be merely a useful fiction. There are difficulties with both options (those with ersatzism are discussed at length in Lewis (1986), those with fictionalism in Rosen (1990) and (1993*b*)). And in any case, ersatzism is incompatible with nominalism, and fictionalism is uncongenial to reconstructivism. A third alternative, more coherent with reconstructive nominalism, would be to reconstrue talk of 'possible worlds' in terms of primitive modal operators. Chihara has reserved detailed discussion of just how this is to be done for a later occasion, and it can be expected to become the topic of his third book. Even in the second book, Chihara already gives considerable space intermittently throughout the first half of the volume to discussion of philosophical criticisms of modality in general and primitive modal operators in particular of the kinds merely mentioned in passing in Chapter II.B.

In the second half of his second book, Chihara offers a polemical survey of nominalism and anti-nominalism in the 1970s and 1980s. (He makes unusually extensive use of unpublished remarks, undoubtedly not in all cases representing considered opinions, of the philosophers he criticizes.) Of authors we have cited or will be citing, Field is the target of one whole chapter and the larger part of another, the nominalist Kitcher and the realist Maddy are each separately the target of a chapter, while the structuralists Shapiro and Resnik are jointly the target of another. (Also the target of a chapter is Burgess (1983), and the criticism there is one we have tried to take into account in our Conclusion.) In his concluding chapter, Chihara addresses the distinction between hermeneutic and revolutionary nominalism, though without adopting that jargon. On the one hand, he is dissatisfied with the position of those rival nominalists whom he takes to have been claiming that their reconstruals give analyses of what ordinary mathematical assertions mean, and makes no such claims for his own reconstrual. On the other hand, he is dissatisfied with the position of those rival nominalists whom he takes to be simply declaring ordinary mathematical assertions to be false, without attempting any reconstrual under which they would be true. He takes Kitcher and Field as his main examples of these two types. Chihara describes his own judicious position towards the axioms of set theory as being that, while he does not accept the axioms as 'literally' true, he holds that 'there is much truth in' set theory. The exact meaning of this formulation may be somewhat elusive.

c. *Bostock, Hodes, and Bigelow* et al.

There are several proposed reconstruals of classical mathematics in the literature that have at least the following in common: first, all draw to some degree or other on one strand or other of the logicist tradition; second, all invoke modality, if in no other way then at least to the extent of assuming the possible existence of infinitely many individuals; and third, all reconstrue talk of mathematicalia into talk of entities somewhere below sets on the scale of abstractness that leads downwards from sets through attributes to formulas as abstract types and on to formulas as concrete tokens. Chihara's proposal is the only one that gets all the way down to the bottom of the scale, rather than stopping at some intermediate stage. Some others among these proposals may be mentioned, since their techniques may be adaptable for purposes of nominalistic reconstrual; but they will be mentioned only briefly, since they are not themselves nominalistic reconstruals.

David Bostock, in a two-volume work (Bostock 1974–9), carries out a detailed reduction of mathematicalia to Russellian 'propositions'. The preface to his second volume suggests that his motivation is nominalistic, and that for him talk of 'propositions' should eventually be further reduced to talk of possible tokens, or rather, of possible acts of tokening, 'possible sayings or thinkings'. He does not, however, carry out such a further reduction, or discuss the nature of modal logic required.

Harold Hodes followed up a position paper (Hodes 1984*b**) with long technical papers (1990, 1991) carrying out a reduction of numbers and sets to Fregean 'concepts'. (His technical work on modal logic cited in article II.B.3.b seems to have been partly motivated by its connection with this strategy.) He indicates that he considers himself, in so doing, to have replaced 'heavy' ontological commitments by 'light', and that he considers his position is not 'realist' but 'conceptualist'. That is to say, he thinks the 'concepts' are immune to the kind of quasi-nominalistical arguments (alluded to in article I.A.2.d) he advances against numbers and sets, and perhaps takes (facts about) 'concepts' to be somehow derivative from (facts about) something more linguistic and more concrete. He does not, however, undertake any explicit reduction of the theory of 'concepts' to a theory of more concrete entities. Hodes seems to think of himself as giving an analysis of what ordinary mathematical assertions mean.

John Bigelow, in his book Bigelow (1988), and his subsequent contribution Bigelow (1990)* to the Irvine volume, carries out a reduction of mathematicalia to universals, conceived as in Armstrong (1978). He declares himself, like David Armstrong, a 'realist' about universals, and proposes no further reduction of them. He also, however, declares himself a 'physicalist', and in this respect he has a certain sympathy with some of the underlying motivation of some nominalists, despite his rejection of nominalism itself. For him, **physicalism** is the doctrine that the only entities are physical particulars and physical universals. Given the relationship that should hold between predicates in an acceptable language and universals, it would seem to follow that the only acceptable predicates are physical predicates.

Bigelow cites as 'broadcasting on nearby wavelengths' several other proposals suggesting some moderate physicalistic reconstrual of mathematics:

> his collaborator Robert Pargetter and other members of Armstrong's school, including Armstrong himself (as in Armstrong 1991);

> the self-described 'realist' or 'physicalistic Platonist' Penelope Maddy (as cited in articles I.A.1.b and I.A.2.a, and as represented in the Irvine volume by Maddy 1990*b*);

> the self-described 'structuralists' Stewart Shapiro, as in Shapiro (1983*a*), and Michael Resnik, as in Resnik (1981, 1982) (also represented in the Irvine volume, by Resnik 1990).

These positions can at most be described as hemi-demi-semi-nominalistic, since all the writers named have been notable critics of full-fledged nominalism (and especially of Field, as indicated in article 1.b). (For some difficulties with the background metaphysics assumed by Armstrong's school and its applications in the philosophy of mathematics, see Rosen (1995).)

d. Putnam₁₉₆₇

Putnam, author of the book Putnam (1971) most often cited for the claim that abstract entities cannot be dispensed with in mathematics and science, was also author of a paper (Putnam 1967*) suggesting how they

could be, using modal logic. The technical details of the modal reconstrual he proposed are of no continuing interest, among other reasons
because he did not deal with mixed, mathematico-physical language;
but the apparent contradiction in his position in two works separated by
only a few years requires comment. Putnam is (like Russell) a philosopher
notorious for frequent radical changes of view, so that his students and
others have taken to adding a subscript date to his name when referring
to him; and it is entirely conceivable that Putnam$_{1971}$ really does contradict
Putnam$_{1967}$. Close reading, however, suggests the position(s) in the two
works cited are compatible.

Putnam$_{1967}$ did not call the reconstructed theory in his paper 'nominalistic' because he understood 'nominalism' as including the rejection of modality as well as of abstracta. Moreover, even if he had called
the reconstructed theory 'nominalistic', he would not have called himself a 'nominalist', owing to his attitude towards the relationship between
the reconstructed and the original theory. He calls them **equivalent
descriptions**. While the positive content of this label is not entirely clear,
he does clearly repudiate both the hermeneutic view that the reconstructed theory can be regarded as an analysis of what the original theory
really meant all along (or vice versa), and the revolutionary view that the
reconstructed theory is a distinct and better theory to be believed instead
of the original (or vice versa). Putnam$_{1971}$ did not repudiate this doctrine
of 'equivalent descriptions' in his book, but rather listed it in his last
chapter among topics there was not space to discuss. Moreover, when
in the book he compares classical mathematics to 'nominalism', calling
the latter inadequate for science, he clearly uses 'nominalism' for the
original overall position of Goodman and Quine, which involved not only
nominalism in the proper sense of rejection of abstracta, but also rejection of modality. Often, what he compares classical mathematics to is not
'nominalism' at all, but constructivism or intuitionism or predicativism.
It may very well be, then, that if he had had the space to discuss 'equivalent descriptions', then rather than say that abstract entities are indispensable for science, he would instead have said that classical mathematics,
either in its usual version with abstracta, or in an 'equivalent description'
with modality, is indispensable.

e. Hellman [cf. article III.A.1.c, Chapter II.C]

Geoffrey Hellman had a double aim in his book Hellman (1989). On
the one hand, he wished to advocate modal structuralism—inspired by

Benacerraf (1965) in its structuralist aspects, and Putnam (1967) in its modal aspects—and he seems to have wished to advocate it as an analysis of what ordinary mathematical assertions mean. On the other hand, he wished to suggest how a modal structuralist reconstrual could be pressed further to provide a nominalistic reconstrual. It is in this sense that Chapter II.C of this book is inspired by Hellman's work. A convenient summary of his position, with some afterthoughts, is available in his contribution to the Irvine volume (Hellman 1990*): it suggests a deeper commitment to modal structuralism than to nominalism as such. (His later work has been more concerned with constructivism and intuitionism and predicativism than with nominalism.) In technical details, Hellman did not make use of plethynticology, not having available the work of Lewis (1993)* and co-workers drawn on in Chapter II.C. Moreover, the modal logic he used was different, so that article A.1.c is closest to being a summary of his approach. (The proximate source for Chapters II.B, II.C was Burgess (1995).)

3. MISCELLANEOUS STRATEGIES IN THE LITERATURE

a. Kitcher [cf. article A.1.e]

Philip Kitcher presents a self-consciously psychologistic and historicist approach to the epistemology of mathematics. He holds that any such approach must also be nominalistic, and devotes a chapter of his book, Kitcher (1984: chapter 6)*, to a nominalistic reconstrual of mathematics. In line with his psychologistic and historicist orientation, he seems to hold that his reconstrual provides an analysis of what ordinary mathematical assertions mean. This is a claim that may seem very surprising in view of the nature of the reconstrual; for it is article A.1.e that provides a summary of Kitcher's approach.

In making comparisons, it should be noted that the terminology used in Kitcher's book differs from that used in article A.1.e: he writes of 'ideal agents' rather than 'superhuman minds', and while a mind was taken in section A.1.e to be a bodiless and hence presumably sexless 'it', for Kitcher an agent is always a 'she'. Further—and this is why Kitcher is placed here rather than in the preceding section 2 on modal nominalism—it should be noted that for Kitcher a modal formulation like that given in article A.1.e is at best a colloquial stand-in for a more elaborate formulation involving his account of the nature of 'idealizing' theories.

Very roughly, what corresponds on that account to an assertion about what a hypothetical entity, if it had existed, necessarily would have done, is an assertion about what follows from the stipulation of what an ideal entity is like. Equally roughly, what corresponds to an assertion that the hypothetical entity could have existed is not, as one might expect, an assertion that the stipulation is suitably coherent. To take that line would leave open the possibility of obtaining a priori knowledge of mathematics, which it is the aim of Kitcher's psychologistic and historicist approach to close off. Rather, what corresponds to an assertion that the hypothetical entity could have existed is an assertion that the idealization is 'appropriately grounded' or is 'relevantly approximated' by real entities.

The approximation need not be very close: it is just we human beings who are supposed to 'ground' the supreme Agent. Our all-too-finite, all-too-fallible collecting abilities are supposed to 'approximate' Her infinite, infallible ones. But Kitcher's account of idealizing theories defies brief summary, and in this sense so does his reconstrual of mathematics as She'ite theology.

b. Gottlieb and Bonevac [cf. article A.2.a]

Substitutional logic of one kind or another has been the main or sole logical device in at least two projects for nominalistic reconstrual of mathematics. The first such strategy is that in a book (Gottlieb 1980) by Goodman's student Dale Gottlieb. Like his teacher, Gottlieb appeals to 'philosophical intuition', and in fact devotes the opening pages of his book to listing a whole series of such intuitions. Only a tiny part of elementary arithmetic is reconstrued by his strategy.

The second such strategy is that in a series of papers (Bonevac 1983, 1984, 1985) of Daniel Bonevac, following up on his book Bonevac (1982) on the general issue of the aims and claims of reduction in the abstract sciences. The whole of set theory is claimed to be reconstruable on this strategy. In this sense the simple strategy in article A.2.a of this book is loosely inspired by (or is something of a parody of) Bonevac's more complex approach. On the question of the status of substitutional logic, Bonevac sometimes seems to suggest that 'ordinary language quantification [not some exceptional or idiomatic cases, but in regular and paradigmatic cases] is substitutional', and sometimes claims that 'substitutional quantification involves no ontological commitment'. Together these claims or suggestions would imply that ordinary language quantification involves no ontological commitment. There is a difficulty here, discussed in the penultimate section of the study Kripke (1976) of substitutional quantification. For 'ontological commitment' is a technical term, introduced by a stipulative definition, according to which, nearly enough, ontological commitment just *is* that which ordinary language quantification, in regular and paradigmatic cases, expresses.

c. Unnamed [cf. article A.2.b]

Predicate functor logic has not yet been deployed by nominalists hoping to eliminate numbers and sets, but as Johan van Benthem observes in an interesting paper (van Benthem 1977) on the debate over primitivism in tense and modal logic, something very like it has been deployed by those who would like to achieve the full expressive power that quantifying over instants and worlds provides, while avoiding the assumption of the existence of such entities. They have tended to propose adding new tense and modal operators that have no obvious reading in English (or at least none not involving such phrases as 'there will have been a time' or 'there is a contingency'), and taking them as primitives. What the logicians of this type were thus doing, according to van Benthem, was introducing one by one the very kind of functors that C. S. Peirce and other successors of George Boole resorted to in attempting to progress beyond traditional term logic and the syllogistic. Their work, along with later attempts of Jan Łukasiewicz, Leśniewski, and others, to rehabilitate traditional term logic by extending it beyond the syllogistic, provides the historical antecedents of Quine's predicate functor logic. The temporal and modal primitivists cited by van Benthem have thus, so to speak, been unconscious advocates of functorial logic. Because they have not been conscious advocates, their names have not been put at the head of this article.

d. Routley (a.k.a. Sylvan) [cf. article A.2.c]

Richard Sylvan (né Routley) has advocated an idiosyncratic version of diacritical and dialectical logic as a solution to the problem of the abstract entities—and to most other problems of philosophy as well. An entertaining and enlightening account of his 'noneism' can be found in Lewis (1990).

C
Conclusion

0. OVERVIEW

The title at the head of the following remarks should really be 'In Lieu of Conclusion'. For the remarks on the significance of reconstructive nominalism to follow will not be conclusions drawn from anything established in earlier chapters; nor will they be conclusive. We will confine ourselves to making a few remarks on two questions that were enunciated in our Introduction, but whose consideration was there postponed until after the presentation of the various reconstructive strategies.

The question we take up in section 1 is that raised at the very end of our Introduction (in section I.A.3), namely, the question of the scientific merits of a nominalistic reconstruction as an alternative to or emendation of current physical or mathematical theory. This is a question only for those who profess to adopt a naturalized rather than an alienated epistemology, and in particular only for those reconstructive nominalists who profess to be proposing an internal revolution in science, not an external invasion of science by philosophy. The question of what non-, un-, or anti-scientific philosophical merits might be claimed for a nominalistic reconstruction from a standpoint prepared to appeal outside, above, and beyond scientific standards of merit to some supposed extra-, supra-, preter-scientific philosophical standards—to the Oracle of Philosophy or to occult faculties of 'philosophical intuition that cannot be justified by appeal to anything more fundamental'—will not concern us.

Two observations immediately suggest themselves. First, the question of the scientific merits of nominalistic reconstruction is really many questions, one for each reconstructive strategy surveyed in Part II and Chapters A and B of this part. For presumably a reconstruction involving point-instants of space-time and one involving the possibility of constructing linguistic expressions, for instance, cannot both be scientifically optimal. Our remarks, however, will be of a general character, turning on features common to the theories produced by most or all of the various nominalistic strategies: apparent great economy of abstract ontology;

omission of substantial parts of pure mathematics; admitted impracticality for much scientific work; and so on.

Second, the question is really not ours as philosophers to answer. For ultimately the judgement on the scientific merits of a theory must be made by the scientific community: the true test would be to send in the nominalistic reconstruction to a mathematics or physics journal, and see whether it is published, and if so how it is received. This, however, is a test to which reconstructive nominalists have been unwilling to submit (and prudently so, in view of the omission and impracticality just alluded to). And that raises a rather delicate question, to which we devote article 1.a below (following up on earlier remarks in Rosen (1992: chapter 5)): in what sense can philosophers proposing a revision of science claim to be judging by scientific standards, if they will not leave the merits of their proposal to be judged by practising scientists?

We do not pretend to provide a complete answer, but we do think the claim to be judging by scientific standards will be quite untenable unless the most conspicuous feature and alleged merit of the nominalistic reconstructions, their economy of abstract ontology, counts as a scientific merit. And that raises a further question, to which we devote article 1.b below (following up on earlier remarks in Burgess (1990b)): if economy of abstract ontology does count as a scientific merit, then one would expect to find somewhere in the historical record of the period since natural science first separated itself from natural philosophy instances of workers in the empirical and/or mathematical sciences showing an attraction to theories that were conspicuously economical and/or an aversion to theories that were conspicuously extravagant in abstract ontology; but is this what one does find? Has this been what was at issue in any important case in the history of science where there has been hesitation and controversy in choosing between two theories?

This again might be thought a question that is not really ours as philosophers to answer, a question that should be left to professional historians of science. And indeed, it is only because the professionals have not much examined it that we admitted amateurs presume to do so. In doing so, we do not pretend to provide a conclusive answer, but at most some suggestive observations. What we suggest is that it is at least very difficult to find any unequivocal historical or other evidence of the importance of economy of abstract ontology as a scientific standard for the evaluation of theories.

The question we take up in section 2 is the question, mentioned early in our Introduction (in article I.A.o.a) but in line with the policy there ɛ nnounced systematically ignored ever since, of the status of hermeneutic nominalism, of the merits of nominalistic reconstruals as linguistic or semantic analyses or exegeses of the meaning or sense of standard scientific language. Again this is really many questions. For a modal structuralist analysis and an ideal agent analysis, for instance, cannot both be faithful accounts of what standard language means. Again, however, our remarks, though illustrated by the example of one specific reconstrual, will turn on a feature common to them all: the fact that each is produced by a systematic method of paraphrase that preserves all apparent logical implications.

Again the question seems one that it is not for us as philosophers to answer. The question of what evidence there is to favour any one hermeneutic hypothesis over any other (or over the null hypothesis that 'deep down' standard scientific language really means just about what it appears to mean 'on the surface') seems one best left to professional linguists without ulterior ontological motives. And indeed, though we find all the analyses and exegeses very implausible as accounts of the 'sense' or 'meaning' of standard language (at least in any sense or meaning of 'sense' or 'meaning' having anything to do with speakers' and writers' intentions or hearers' and readers' understandings), we are prepared to leave that issue to the linguists. Our discussion will focus not on the evidence for this, that, or the other hermeneutic claim, but rather on the relevance of any hermeneutic claim, supposing for the sake of argument that its correctness is granted.

A hermeneutic claim is supposed to be relevant because of the compatibility claim that is supposed to follow from it, namely the claim that nominalists, in denying for instance that there are any such things as numbers, are not thereby contradicting anything asserted by physicists or mathematicians when they make assertions for instance about 'Avogadro's number' or 'the Bernoulli numbers'. What we question (in article 2.b, after an optional quasi-historical digression in article 2.a on a bit of philosophical jargon whose use has been the source of much obscurity and confusion in this area) is whether this compatibility claim does follow. What we suggest is that it is at least very difficult to see how it could follow.

It is obvious even from the foregoing rough, brief account of the direction our remarks will be taking that we will not be considering every kind of claim that might be made on behalf of every kind of nominalistic strategy. That is one reason why our discussion can at most claim to be suggestive. What we think it suggests is that there are considerable grounds for doubt about whether nominalistic reconstruals really are significant in the types of ways reconstructive nominalists often seem to take them to be—enough so, at any rate, to make it worthwhile to consider what significance might be claimed for a nominalistic reconstruction from a non-, un-, or anti-nominalist viewpoint. If nominalistic reconstruals are not plausible as analysis of the ordinary meaning of scientific language, and if nominalistic reconstructions are not attractive by our scientific standards as alternatives to current physical or mathematical theories—if nominalism makes no contribution to linguistic science, nor to physical or mathematical science—then must the programme of nominalistic reconstrual be judged an academic exercise in the pejorative sense of the term, an intellectual entertainment addressed to no serious purpose? This is a question that for obvious reasons is seldom considered by those engaged in developing strategies of nominalistic reconstrual. It is the question we take up in the closing section 3.

1. THE RELEVANCE OF REVOLUTIONARY NOMINALISM

a. *Reconstruction and 'Occam's Razor'*

It is customary to contrast theory and practice, but scientific theorizing is itself a practice, namely, a practice of accrediting theories, where the theory in a given area that scientists credit is the one that they use in certain ways, for instance, as a source for the general principles deployed when giving the most considered theoretical explanations, for the background assumptions taken for granted when testing novel hypotheses, for the standards of comparison invoked when judging whether simplifying assumptions facilitating computation will still give an approximation good enough to be reliable for purposes of technological applications, and so on. Like any activity and practice of choice and accreditation, science embodies certain standards for choice, certain norms for accreditation. Becoming a participant in the practice, becoming a scientist, involves learning how to distinguish between more and less credible theories, and thus involves acquiring knowledge of the norms.

But there is no official book of rules for the direction of the scientific mind. The active, practising scientist's knowledge of the relevant standards and norms is largely tacit or implicit, rather like speakers' knowledge of the rules of grammar of their native language. The task of making explicit what scientific norms are belongs to **descriptive methodology**, a branch of naturalized epistemology. It contrasts with presuming to dictate what scientific norms ought to be, which constitutes **prescriptive methodology**, a branch of alienated epistemology. The two stand to each other rather as the descriptive grammar of Chomsky stands to the prescriptive grammar of Fowler.

There is a fair degree of agreement among descriptive methodologists on a somewhat heterogeneous list of features that tend to be implicitly used in science as standards for judging when a theory is a better choice than its rivals:

(i)   correctness and accuracy of observable predictions
(ii)   precision of those predictions and breadth of the range of phenomena for which such predictions are forthcoming, or more generally, of interesting questions for which answers are forthcoming
(iii)   internal rigour and consistency or coherence
(iv)   minimality or economy of assumptions in various respects
(v)   consistency or coherence with familiar, established theories, or where these must be amended, minimality of the amendment
(vi)   perspicuity of the basic notions and assumptions
(vii)   fruitfulness, or capacity for being extended to answer new questions

Our numbering these features, as if they were Seven Cardinal Virtues, is not to be taken seriously. Most of the items listed are less single features than clusters of features; some overlap; one may often be in tension with another; there is no fixed priority among them. The list is a long way from an explicit representation of the tacit knowledge or skill possessed by a competent working scientist. Still, it is useful to the extent that it identifies some of the terrain upon which contests are to be fought by philosophers professing allegiance to naturalization or naturalism in epistemology.

One way such a philosopher might try to judge whether the nominalistic reconstructions are scientifically so superior that they ought to be credited in preference to current theories (or at least scientifically so meritorious that current theories ought no longer to be fully credited in preference to them), would be to examine the features of the nominalistic reconstructions and compare them to a list of recognized scientific merits like that just given. However, the data from which such lists are compiled come from the record of judgements of the scientific community as to which theories to credit, much as the data from which lists of grammatical rules are compiled come from the record of the judgements of native speakers about what utterances are grammatical. No list, however widely agreed among descriptive methodologists, has more authority than the data. (An individual choice by scientists may be judged anomalous, as may a particular utterance by a native speaker; but in either case only because the item in question does not fit with the pattern of the rest of the data.)

So a far more direct test would simply be to submit a paper presenting the nominalistic reconstruction to, say, the *Physical Review*, and gauge the reaction. This, however, is a test to which, so far as we know, none of the nominalists whose work we have surveyed has subjected himself. (At any rate, no papers on nominalistic physics have appeared in the journal mentioned.) And it is easy to understand why, since it is easy to imagine what the reaction to the submission of such a paper to such a journal would be. Contrast two imagined innovations in science. On the one hand, imagine that it were discovered how to understand the present system of fundamental particles (a half-dozen species of quark, each with its anti-quark, and a comparable number of leptons, each also with its anti-particle) as arising from combinations of just two or three ultra-fundamental particles (hyper-quarks, if you will). On the other hand, imagine that it is discovered that by a certain judicious choice of logical devices it is possible to frame a version of the present theory of fundamental particles that avoids involving numbers, functions, or sets. The first discovery would be front-page news, in the *Physical Review* if not in the *New York Post*. The second would be received rather differently, we suspect, as a curiosity if not simply as a particularly clumsy notational variant of current theory.

The innovation, we suspect, simply would not be recognized as progress by practising scientists. And this is so not just for physics, we suspect, but for every natural or social science. Would it be reckoned a significant advance in evolutionary biology to show that all reference to species could be avoided in favour of a complex idiom that countenanced only individual organisms? Would economics profit by its own lights from the demonstration that reference to choice functions and indifference curves could be replaced by quantification over material objects of some unobvious sort? In terms of the list of standards given above, the reconstructive nominalist seems to be giving far more weight to factor (iv), economy, or more precisely, to a specific variety thereof, economy of abstract ontology, than do working scientists. And the reconstructive nominalist seems to be giving far less weight to factors (v) and (vi), familiarity and perspicuity.

These factors are very important in scientific work. They make for ease of use of the theory in the several ways mentioned above, for ease in communicating and testing the theory, and in revising it if necessary and extending it if possible. It should be stressed that this is something that matters not just to technicians and engineers, but also to even the most pure and theoretical of pure and theoretical scientists. Certainly it would be madness to suggest that applied physicists or economists interested in predicting the perturbations of Mars before a space-shot or the fluctuations of the peso before an intervention by the Central Bank should carry out their reasoning in the language of synthetic geometry or modal logic rather than of mathematical analysis. But it would hardly be more sane to suggest that purely theoretical explanations of these astronomical and financial phenomena, or proposals to refine existing gravitational or monetary theory, should be couched in synthetic or modal language.

Moreover, the features (v) and (vi) are closely connected with feature (vii), fruitfulness. The power and flexibility of the apparatus made available to the empirical sciences by modern mathematics permits a theory to be recast in various ways, some amenable to generalization or amendment in some directions, others in others. Having a variety of formulations available is of considerable importance both when one wants to extend a successful theory and when one needs to patch up a theory that has run into difficulties. It is conceivable that having a synthetic geometrical version or a modal logical version available in addition to more orthodox versions might have some utility; it is virtually certain that discarding the orthodox versions, leaving only a synthetic geometrical or modal logical version available, would have much disutility.

Indeed, to give a high score to any of the nominalistic reconstructions would require one to discount these factors (v)–(vii) almost entirely. Thus it would seem that if the nominalistic reformulations are to be claimed to be of superior merit by the standards of the sciences, familiarity and perspicuity and fruitfulness must somehow be expunged from the list of scientific 'merits'. And more obviously, the mathematical sciences—so often considered by non-philosophers the very model of a progressive and brilliantly successful cognitive endeavour—must somehow be expelled from the circle of 'sciences'. For with a few exceptions, the nominalistic strategies we have surveyed simply discard whatever of pure mathematics has not yet found application in the empirical sciences, or at least whatever goes beyond classical analysis. It would seem that if the nominalistic reformer is to claim to be an adherent of naturalization or naturalism in epistemology, the 'naturalism' in question must be of a restricted variety, making invidious distinctions, marginalizing some sciences (the mathematical) and privileging others (the empirical). And it must also be of a selective variety, again making invidious distinctions, and down-playing the importance of some norms (familiarity and perspicuity and fruitfulness) while playing up that of others (economy or parsimony, specifically or especially of abstract ontology).

Now this is not 'naturalism' as exemplified by such otherwise diverse anti-nominalists as Maddy and Lewis, and before them Gödel and Carnap (as quoted or cited in article I.A.2.a), and it is indeed quite unclear how a nominalist professing 'naturalism' could justify abridging the roll of sciences, or expurgating the list of scientific merits. Recall that naturalism (as we introduced it in article I.A.2.a) was supposed to derive from a partial acceptance and partial rejection of empiricism and scepticism. On the one hand, naturalism was supposed to accept the traditional empiricist claims that science goes far beyond anything directly supported by the evidence of the senses, and that in so doing it is guided to a significant extent by features that reflect our practical limitations and natural proclivities more than they reflect 'reality as it is in itself'. On the other hand, naturalism was supposed to reject traditional sceptical claims that beliefs that go beyond what is directly supported by and founded on the evidence, or that are guided or shaped by our limitations or proclivities, must necessarily be unjustified.

By contrast, the epistemological stance of the reconstructive nominalists begins to look like a questionably coherent combination of naturalistic and traditional, pre-naturalistic elements. There is enough of naturalistic deference to science to close off the easy route of simply classifying science or applied mathematics as 'useful fiction' (the option discussed in section I.A.3). But there is enough left of the view that only what is supported by the evidence of the senses can be genuine knowledge to lead to the dismissal of pure mathematics, and enough left of the view that what is shaped by our practical limitations is not authentic cognition to lead to the neglect of familiarity, perspicuity, and fruitfulness as standards for the choice of theory. It is needless for us to say that a thoroughgoing sceptical empiricist would find claims to knowledge of a continuum of point-events of space-time, or of a plethora of unactualized possibilities, no more acceptable than claims to knowledge of a realm of abstracta. What we want to add is that a thoroughgoing naturalist would take the fact that abstracta are customary and convenient for the mathematical (as well as other) sciences to be sufficient to warrant acquiescing in their existence. Reconstructive nominalists professing 'naturalism' begin to seem neither fish nor fowl. Thoroughgoing sceptics and thoroughgoing naturalists alike would want to ask them the question that agnostics and

fundamentalists alike tend to ask of those liberal theologians who acknowledge the authority of scripture in a general way, but stick at some of its
more primitive or more demanding precepts: with what right do you pick
what you pick and neglect what you neglect, given that the two sorts of
principles enjoy the same connection to the one source of authority you
profess to acknowledge?

A possible response might be that science itself makes invidious distinctions. Some of its branches are considered more speculative and con-
jectural than others. Perhaps also some of its standards are considered
'merely pragmatic' as opposed to 'genuinely cognitive' virtues. A related
response would be that science itself observes a division of labour. For
example, it is generally conceded that mathematically rigorous theories
are scientifically superior, and yet it is equally generally conceded that
theoretical physicists need not give rigour their attention in the short run.
And in fact, theoretical physicists leave most issues pertaining to rigour to
mathematical physicists, who like other applied mathematicians leave
many such issues to pure mathematicians, who in turn leave some such
issues to specialists in mathematical logic. (Developments in logic do
filter back into pure mathematics; developments in pure mathematics do
filter back into mathematical physics and other branches of applied mathematics; developments in mathematical physics do filter back into theoretical physics—but all this only in the long run.) There may well be
much to be said for a line of response something like this. However, the
reconstructive nominalists who profess 'naturalism' have not themselves
much said it. They have not proceeded by first presenting studies of the
distinctions and divisions observed within the community of working
scientists, and then citing these as warrant for discarding pure mathematics and ignoring familiarity, perspicuity, and fruitfulness. So one may
well ask what the source of their warrant is supposed to be.

There is, however, no mystery here. The reconstructive nominalists
who profess 'naturalism' generally present themselves as responding to
'the Quine–Putnam indispensability argument', and that anti-nominalist
argument makes the major concession to nominalism that it is only indispensability in principle (not in practice) and indispensability for empirical
(not mathematical) science that counts. (The latter point, at least, is tell-
ingly made by Charles Parsons in Parsons (1986) and elsewhere.) This, of
course, only pushes back the question: why were professed anti-nominalists
so willing to make such concessions to the opposition? Well, perhaps they
did so 'just for the sake of argument'.

Perhaps we should do so 'just for the sake of argument' ourselves. Let us grant the negative side of the professedly 'naturalist' reconstructive nominalists' claim: that mathematics doesn't count as a science, and that familiarity, perspicuity, and fruitfulness don't count as standards. Still, one may question their positive claim, that simplicity or economy of the kind exhibited by their proposed alternatives to current scientific theories, that simplicity or parsimony specifically or especially of abstract ontology, does count as a scientific merit. This is something the proponents of the 'Quine–Putnam indispensability argument' seem never to have questioned. And yet it seems to us by no means obvious, and a point calling for investigation.

It will perhaps be well to begin our investigations, to which we devote article 1.b below, with 'Occam's Razor'. For something called 'Occam's Razor' is much mentioned by nominalists, and something called 'Occam's Razor' is much mentioned in discussions of scientific method.

b. Occam's Razor

A version of 'Occam's Razor' is often alluded to in Martin Gardner's books debunking pseudo-science. On the cover of the paperback edition of one such book, Gardner (1989), the publisher quotes a reviewer of the hardcover edition to the effect that Gardner 'wields Occam's Razor like a switchblade'. Just what is the weapon thus wielded? A typical article in the book combines a little edifying discussion of why especial caution is needed in testing claims about the occurrence of para-normal phenomena with a lot of entertaining reportage about how the requisite caution conspicuously failed to be maintained in this, that, or the other highly publicized case. 'Occam's Razor' is the label used in alluding to the maxim of scientific method that enjoins especial caution in such cases. But just what maxim is that?

Exact formulation of the maxim is not Gardner's highest priority. If he were only concerned with nineteenth-century 'spiritualism', which explicitly claimed disincarnate spirits were at work in the material world, a fairly simple formulation would do. Whatever the status of 'physicalism' as variously understood by philosophers, there is now a kind of **physicalism** widely, if implicitly, accepted among scientists, enjoining caution about the claims of spirit mediums and their ilk:

(i)   aphysical agents are not to be posited unless necessary:
        explanations in terms of aphysical agents
          are not to be resorted to until
        explanations in terms of physical agents have been exhausted

Twentieth-century para-psychology is much less definite, much more fluid, about its claims. In para-psychology, sometimes the suggestion seems to be less that certain aphysical agents are at work than that certain previously unrecognized sorts of physical agents are. A maxim relevant to such a suggestion would be:

(ii)   physical agents are not to be multiplied beyond necessity:
        explanations in terms of extraordinary physical agents
            are not to be resorted to until
        explanations in terms of ordinary physical agents
            have been exhausted

Such a maxim would be pertinent to mainstream as well as to fringe science, and would counsel strict testing of claims of the discovery of a fourth family of quarks (beyond down/up, strange/charmed, bottom/top), for instance. Formulations (i) and (ii) can be merged into a single ontological maxim, enjoining caution in positing previously unrecognized sorts of causal agents, whether physical or aphysical:

(iii)   causal agents are not to be multiplied beyond necessity

In para-psychology, sometimes the suggestion seems to be less that certain previously unrecognized physical agents are at work than that certain previously recognized ones are at work in previously unrecognized ways. Here one might cite an ideological twin to the ontological maxim (iii), enjoining caution in introducing previously unrecognized causal predicates, connoting previously unrecognized kinds of causal activities:

(iv)   causal activities are not to be multiplied beyond necessity

In mainstream science, this would counsel strict testing, for instance, of claims of discovery of a fifth fundamental force (beyond the gravitational, electromagnetic, and strong and weak nuclear). If we understand 'causal agencies' in a sense neutral between ontological and ideological, between sorts of agents and kinds of activities, (iii) and (iv) can be merged into a single formulation:

(v)   causal agencies are not to be multiplied beyond necessity:
        explanations in terms of extraordinary agencies
            are not to be resorted to until
        explanations in terms of ordinary agencies have been exhausted

Something very like (v) seems to be Gardner's preferred version of 'Occam's Razor'. It is an example of a maxim enjoining parsimony whose status as a rule of scientific method is comparatively uncontroversial.

Another such example might be the following maxim:

(vi)   assumptions are not to be multiplied beyond convenience

This maxim is very broad in scope. It applies to abstract entities that do not as much as to concrete entities that do have to coexist and interact within a single spatiotemporal, causal framework. It enjoins caution in ontology, in recognizing new sorts of entities; but it also enjoins caution in ideology, in recognizing new kinds of predicates connoting new kinds of relationships, even among previously recognized sorts of entities. For that matter, it enjoins caution about making new assumptions, even about previously recognized kinds of relationships among previously recognized sorts of entities. If despite this breadth (vi) ought to be comparatively uncontroversial, it is because it is so weak in force. It only makes utility or convenience a necessary condition of acceptability, and so only excludes completely arbitrary and gratuitous assumptions.

But what we hinted in article 1.a that we wished to examine in this section is something somewhat different from (v) and (vi), and more controversial. A maxim that enjoined preference for one of the nominalistic reconstructions surveyed in earlier chapters over current theories in the mathematical and empirical sciences would surely have to be one formulated in terms of indispensability in principle rather than convenience in practice, in terms of what it would be possible to do without rather than of what it would be convenient to do without. For even those who hold the new theories to be superior in principle do not urge that they should in practice be used in the day-to-day work of ordinary scientists. Such a maxim would also have to focus specifically on ontology as opposed to ideology, and on the abstract as opposed to the concrete. For to adopt the new theories would require either undertaking substantial new ideological commitments (modal logic, plural logic, or whatever); or else undertaking substantial new concrete ontological commitments (geometricalia, conglomerates, or whatever); or more likely, it would involve both. The maxim would have to read something like:

(vii)   abstract entities are not to be posited unless necessary

Something very like (vii), rather than the less controversial (v) or (vi), seems to be what nominalists have in mind when they speak of 'Occam's Razor'. The question we hinted in article 1.a we wished to consider in this section was whether this version of 'Occam's Razor' is indeed a rule of scientific method. Do scientists tend to prefer theories that are parsimonious and shun theories that are prodigal in abstract ontology?

Let us consider first the empirical sciences. When one thinks of famous controversies over the reality of this, that, or the other item in those sciences, controversies regularly cited by philosophers of science to illustrate their claims, one may think of cases like that of entelechies and vitalist biology; or one may think of phlogiston, or caloric fluid, or luminiferous ether; or one may think (to name a case where the controversial entities were ultimately accepted) of the case of atoms. These are all, however, clearly cases of dispute over an alleged concrete causal agent, physical or otherwise. Our question is: has abstract ontology ever been what was at issue in any important case of dispute between proponents of rival theories in empirical science? Though nothing conclusive will be established, we think it suggestive to take a closer look in this connection at certain disputes in linguistics (to which we will be returning in section 2).

Though semantics, especially as it was forty or fifty years ago, is hardly anyone's paradigm of a science, it may be well to begin there anyhow in illustrating the kinds of complications that can arise in attempting to determine whether a dispute really was over abstract ontology. Consider, then, the dispute between Quine and Carnap over the existence of meanings. Quine himself has more recently returned to the issue, in his reply to Alston in Hahn and Schilpp (1986):

Hypostasis of meanings is a red herring. I keep urging that we could happily hypostasize meanings if we could admit synonymy. We could simply identify meanings with the classes of synonyms. . . . The point . . . is that the prior assumption of an unexplained domain of objects called meanings is no way to explain synonymy or anything else. Synonymy, not hypostasis, is the rub. Given synonymy, a domain of meanings is trivially forthcoming for whatever good it would do.

The post-nominalist Quine insists that provided the equivalence relation of synonymy can be made sense of, he has no objection to anyone's introducing meanings as the abstract characters that expressions that are equivalent in the sense of being synonymous thereby have in common. His objection is, rather, that the equivalence relation of synonymy has not been made sense of. Moreover, he suggests that this was his crucial objection all along, even during his nominalist phase; and certainly his most famous early attack on meanings, Quine (1951*b*), is based more on behaviourist than on nominalist assumptions.

Here is a case where what at first glance seems to be an issue about abstract ontology at second glance seems to be an issue about ideology, or to be intertwined with an issue of ideology of at least equal scientific importance. Such cases we will call cases of **type A**. In this particular case, more specifically, what presents itself initially as an issue about the existence of a certain kind of equivalence characters turns out subsequently to be an issue about the meaningfulness of a certain kind of equivalence predicate.

Perhaps syntax, even as it is today, will not be anyone's paradigm of a science, either, but it may be well nonetheless to turn to it next, and to disputes between Chomsky and two waves of critics over the reality of his deep structures (and their successors at later stages in the development of his thought). Here again Quine's suspiciousness has outlived his nominalism, and was never wholly based on nominalistic scruples alone. Such scruples, indeed, would have been grounds for objection to surface structures as much as to deep ones, and to the sentences that have the structures as much as to the structures that they have. Again it is really behaviourism that aligned Quine with B. F. Skinner and Leonard Bloomfield in the first wave of Chomsky's opponents.

The minimum that linguists seek to do is to develop a formal grammar that is what Chomsky calls 'descriptively adequate', one that judges grammatical the same sentences that native speakers of the language judge grammatical. In dispute with the first wave of opponents just mentioned, Chomsky maintained that even minimal descriptive adequacy cannot be achieved if linguists are hampered by behaviouristic constraints forbidding them to introduce any theoretical classifications that are not directly correlated with behaviour. The dispute is thus a case of type A, in this respect at least like the dispute with Carnap over meanings.

A second wave of critics, including Scott Soames, as in Soames (1985), have not been proponents of behaviourism or opponents of the introduction of deep structures or the like as parts of abstract models for the description of language. But Chomsky has always maintained that descriptive adequacy is not all that he is trying to achieve, but only the first step towards 'explanatory adequacy'. He has always maintained that one might have two descriptively adequate grammars G and G', one of which was in a significant sense better than the other by being 'internally represented' in speakers' minds and/or brains, there playing a role in the causation of episodes of speaking and/or understanding, writing and/or reading. Chomsky's goal has always been to obtain a grammar that is better in this sense.

Now the assumption that G rather than G' is the grammar that speakers internally represent does not all by itself imply any empirically testable claim about the neurology or behaviour of speakers. One would need

additional hypotheses about how exactly grammars are internally represented neurologically and/or about what exactly the role of these internal
representations is in the causation of verbal behaviour, before the claim
that it is G rather than G' would have any consequences testable by examining speakers' neurology and/or behaviour. But Chomskyan theory includes
no such additional hypotheses. What Chomsky's more recent critics have
insisted is that there would therefore seem to be very little ground for
thinking that the grammars developed by Chomsky and his followers are
any more than descriptively adequate. If the only empirical test to which
these grammars are ever put is the test of agreement with speakers' judgements about grammaticality, then why should we think they nonetheless achieve something more than just such agreement? That is to say,
Chomsky's later critics question whether there is sufficient evidence for
positing in the minds and/or brains of speakers something concrete corresponding to the abstract apparatus of deep structures or the like used in
the description of language.

Here is a case where what at first glance seems to be an issue about
abstract ontology at second glance turns out to be an issue about concrete
ontology, or to be intertwined with an issue of concrete ontology of at
least equal scientific importance. Such cases we will call cases of **type B**.
In this particular case, more specifically, what initially presents itself as an
issue about the existence of an abstract entity subsequently turns out to
be an issue about the existence of a concrete correlate thereof. Overall,
considering both phases of debate, the issue of deep structures might be
said to be of **type AB**, with both A-factors and B-factors present.

The same might be said about another dispute there has already
been occasion to mention for other reasons elsewhere in this book, the
substantivalist vs. relationalist dispute touched on in Chapter II.A. Our
discussion there noted the presence of what we are now calling A-factors,
namely, disagreement over the meaningfulness of 'occurs in the same
place but at a different time' and 'occurs at the same time but in a
different place'. It also noted the presence of what we are now calling
B-factors, namely, the dispute over the existence of point-instances conceived of as causal agents corresponding to the mathematical 'points' of
the mathematical 'spaces' used in modelling physical processes. We cannot think of any convincing example in the empirical sciences that is of
type O for 'Occam', where the 'Razor' in something like version (vii) is
central. So we will turn to consider the mathematical sciences, where
candidate examples might be expected to be more abundant.

Various attitudes might be and have been taken towards the possibility of eliminating other mathematicalia in favour of sets (as outlined in article I.B.1.a): ever since Benacerraf (1965) called attention to the philosophical puzzles in this area, they have been cited by philosophers of many different persuasions, who have drawn from them many different morals—including philosophers of the nominalist persuasion, who have drawn from them nominalistic morals, as in Jubien (1977) and Kitcher (1978). Inevitably, then, we are going to have to consider these puzzles at least briefly. Let us begin by describing a couple of the attitudes that might be and have been taken.

MacZee, who attaches great importance to economy of abstract ontology, supposes that since one is going to be assuming sets anyhow, one should now dispense with the assumption of numbers as traditionally conceived, and transfer the terminologies and notations formerly applied to them to set-theoretic surrogates for them. Choosing one convenient system of surrogates (von Neumann's), MacZee therefore now says:

(viii) $\quad 2 = \{\{ \ \}, \{\{ \ \}\}\}$

MacZed's views are just like MacZee's, except for involving a different choice of surrogates (Zermelo's), so that MacZed now says:

(ix) $\quad 2 = \{\{\{ \ \}\}\}$

Though McWye assumes the full cumulative hierarchy of sets over whatever individuals there may be, it never occurs to McWye, who pays no attention to economy of abstract ontology, to doubt that among those individuals, distinct from all sets, are the natural numbers as traditionally conceived.

These by no means exhaust the possible attitudes, though we need consider no others in the present context. (One other possible attitude, 'structuralism', was mentioned in article I.A.2.d and section II.C.0.) It is rather important for higher mathematics that each branch of mathematics, whether in algebra or analysis or geometry, is so conducted that it is possible to adopt an attitude like MacZee's or MacZed's towards it, and view it as simply a branch of set theory. The reason it is important is roughly as follows. (For some further, related discussion, see Burgess (1992).)

Mathematics is no motley: lower mathematics may seem a potpourri of various more or less independent branches, but higher mathematics interweaves strands drawn from all these branches, and so achieves a lofty unity. The great (if not entirely single-minded) French mathematician N. Bourbaki held that, in recognition of this unity, the subject ought to be spoken of in his language in the singular, as '*la mathématique*', and not (as the Académie française decrees) in the plural, as '*les mathématiques*'. It is crucial for the logical coherence of *la mathématique* that the different results brought in from different branches should each have been derived in a logically cogent manner from the basic principles of its branch, and that the basic principles of the different branches should be logically compatible with each other. So long as each branch is conducted in such a way that it could be viewed as a part of set theory, the logical compatibility of the assumptions of the different branches is assured (assuming, of course, the logical consistency of set theory itself). This may be why Bourbaki, though like most mathematicians of his country little interested in set theory as a subject in its own right, chose it as the organizing framework for his enormous, and enormously influential, encyclopedia, *Éléments de Mathématique*.

The foregoing is one way in which the possibility of taking an attitude like MacZee's or MacZed's is important. In connection with the issue about 'Occam's Razor' raised above, however, what matters is what attitude is actually taken by mathematicians: if we could be confident that mathematicians actually think like MacZee or MacZed, then we would have before us an important example of scientists assigning high value to parsimony of abstract ontology; while if we could be confident that mathematicians actually think like McWye, then we would have before us an important example of the opposite kind. Unfortunately, it is very difficult to be confident one way or the other.

Benacerraf's discussion of these matters began, in effect, with the observation that the difference between MacZee and MacZed would make no difference whatsoever to their work as mathematicians, since equations of type (viii) and (ix), with a number-theoretic term and a set-theoretic term flanking an equals sign, simply are not considered in mainstream mathematics (specialized studies in set theory apart). This applies equally to the difference between both Macs and McWye. The failure of mainstream mathematicians ever to discuss equations of the type in question is perhaps most consonant with the hypothesis that their views resemble McWye's, but against this must be set the fact that in introductory textbooks on general set theory studied by non-specialists the equation (viii) does often appear, while (ix) does not, as if in accordance with MacZee's views.

Perhaps the discrepancy here can be reconciled by noting that the preface to what is probably the most popular such textbook, Halmos (1960), ends with the advice to 'read it, absorb it, and forget it'. Just what is supposed to be forgotten is not made explicit, but it surely cannot be, say, the distinction between countable and uncountable sets, or the equivalence of various formulations of the choice axioms; for the author, Paul Halmos, in his other mathematical works assumes familiarity with such material. What Halmos does not ever refer to again are equations like (viii) or (ix), so perhaps these are what are supposed to be forgotten. A full discussion of these matters would require a book-length work in itself. But we have perhaps said enough already to indicate why we think the phenomenon to whose significance Benacerraf first pointed does not teach any obvious lesson one way or the other about how far mathematicians do or do not value economy of abstract ontology.

Perhaps less equivocal examples should be sought elsewhere, in famous cases where the introduction of new sorts of mathematical entities has encountered resistance only overcome after a long struggle. Cases of this kind can be found at every stage in the historical expansion of the mathematics. The most notable controversies were perhaps those over the status of the various kinds of numbers:

> negatives and imaginaries
> infinitesimals
> transfinite cardinals and ordinals

In each of these cases for a long time a significant segment of the mathematical community resisted accepting the numbers in question as anything more than useful heuristic devices suggesting conjectures that would still have to be substantiated by other means. Was the resistance in any of these cases the result of mathematicians being worried over ontological prodigality? Or can the resistance in each case be explained by mathematicians being worried over something else?

The most obvious 'something else' would be rigour (and therewith consistency): the more that reluctance to accept novel kinds of numbers can be explained as resulting from worries about the loose reasoning (and therewith potential contradiction), the less need there will be to appeal to 'Occam's Razor'. And very strong indications of the importance of unease about rigour are found in each case when one considers the closure of the debate. For in each case, the debate closed when there was developed an account of the numbers in question conforming to the highest standards of rigour prevailing at the time. The story is mostly familiar from standard histories of mathematics like Kline (1972).

In the case of negatives and imaginaries, doubts ceased with the discovery, independently by several workers, of the geometric interpretation of negative and complex numbers now routinely taught in high-school algebra. This put those species of numbers on a par with positive real numbers, which had always been interpreted geometrically (as discussed in Chapter II.A). Gauss, one of the co-discoverers of this interpretation, is said to have remarked that if people from the beginning had said 'forwards, backwards, sideways', instead of 'positive, negative, imaginary', there would never have been any controversy.

In the case of infinitesimals, they were banished by Weierstrass when he put the calculus on a rigorous basis. For instance, the definition of:

$$\lim_{x \to a} f(x) = b$$

was changed from:

$|f(x) - b|$ is infinitesimal whenever $|x - a|$ is infinitesimal

to:

$|f(x) - b|$ can be made as small as desired
by taking $|x - a|$ sufficiently small

which is colloquial for:

for every $\varepsilon > 0$ there is a $\delta > 0$ such that
$|f(x) - b| < \varepsilon$ whenever $|x - a| < \delta$

Infinitesimals (and therewith the old definition of limit) were rehabilitated almost a century later, when Robinson's non-standard analysis (mentioned in article A.1.a) provided a rigorous theory of them.

The case of transfinite numbers, or in other words, of Cantor's set theory, is more complicated. Our suggestion that worries about (rigour and) consistency may have underlain much reluctance to accept the theory may well be viewed with suspicion, given that most commentators nowadays tend to down-play Russell's and related paradoxes. We need therefore to explain just what we are and are not suggesting.

One reason recent commentators have de-emphasized the paradoxes is that they played no prominent part in motivating the most articulate and active opponents of set theory, such as Leopold Kronecker, the great forerunner of constructivism, or L. E. J. Brouwer, the founder of intuitionism, or Hermann Weyl, the pioneer of predicativism. Our concern here, however, is with scientific methodology, and how far its rules can be inferred from the reactions of broad segments of the scientific (in the present instance, the mathematical) community during periods of controversy. From this perspective, the large numbers of mathematicians, inarticulate though they may have been, who felt unease about set theory and related developments, and were therefore willing to give the heterodox a hearing, are more important than the quite small number of mathematicians, who having heard one or another heresiarch preach, actually apostatized. Our suggestion is merely that the paradoxes—and the lack of rigour they exposed—contributed significantly to a widespread sense of unease.

Another reason recent commentators have de-emphasized the paradoxes is that they in fact arise only in connection with Gottlob Frege's logical notion of extension or class, not Georg Cantor's mathematical notion of set. Cantor never assumed that every condition determines a set, and indeed his assumptions about when a condition does and when it does not determine a set, as expressed in correspondence with Richard Dedekind (available in Noether and Cavaillès (1937)), in many ways anticipate the axiomatization of Ernst Zermelo. Zermelo's axioms, moreover, are not just a list of assumptions sufficient to derive the main results of Cantor's theory while avoiding the known paradoxes, but are partial descriptions of an intuitive picture, the cumulative hierarchy (expounded in Zermelo (1930)), which gives a fairly strong intuitive conviction of their consistency. Or so it is maintained. We have no wish to deny any of this, but we do wish to point out that the cited publications from the 1930s were not available to mathematicians who had to make up their minds about set theory circa 1900 on the strength of Cantor's main works, or even circa 1910 on the basis of these plus Zermelo's earliest papers. Our suggestion is merely that to such mathematicians it may well have seemed much more difficult than it does to us to distinguish Cantor's notion of set from the inconsistent notion, or to discern in Zermelo's axioms more than an *ad hoc* list.

The tendency to down-play the paradoxes is in large part a reaction against an earlier tendency to over-dramatize and speak of a 'crisis in (the foundations of) mathematics', against which tendency Bernays (1935) already protests. Now indeed, no matter how dire the situation of set theory may have been in the immediate wake of the paradoxes, there was no general crisis in mathematics, if only because set theory did not yet occupy the position of central, organizing framework it was subsequently to be given by Bourbaki and others. Our suggestion is merely that set theory would never have been given this role—most emphatically not by Bourbaki, one of whose most important aims was to raise standards of rigour—until developed with sufficient rigour to show just how the orthodox notion of set differs from the inconsistent notion of class, just which conditions are taken to determine sets, and so on. It was these developments that brought controversy over set theory to a close, so far as the broad mathematical community was concerned (even though, so far as ontology is concerned, the tendency of these developments was if anything to make more conspicuous just how vast the universe posited by set theory is supposed to be). Or so we suggest.

Issues about rigour, we suggest, cloud virtually every case of apparent 'ontological' debate in the mathematical sciences, just as what we called A-factors and B-factors cloud such debate in the empirical sciences. We know of no clear example of striving after economy of abstract ontology in any domain of science, and we are dubious that there is one. But we do not claim expert knowledge here, and leave the matter to the reflection of the informed reader—and in the end, to the judgement of the scientific community.

2. THE RELEVANCE OF HERMENEUTIC NOMINALISM

a. *'Ontological Commitment'*

Before considering more recent views about whether or not standard mathematical and scientific theories imply or presuppose the existence of mathematical and other abstract entities, it may be helpful to review Quine's views on such questions. (It is only helpful, not indispensable; this section is an optional quasi-historical digression, and the reader may skip ahead to article 2.b.) For most more recent discussions still employ terminology and jargon introduced by Quine, and many are quite self-consciously reactions against or responses to Quine's position.

Mathematicians often make assertions like the following:

(i)   The number π is transcendental.
(ii)   There is a number between 10^{100} and $2 \cdot 10^{100}$ that is prime.
(iii)   The number of points of intersection of two algebraic curves is the same as the number of roots of the difference of their defining equations.

But unless they happen to belong to the small minority of mathematicians who are also philosophers, they never explicitly address the question:

Are there (such things as) numbers?

Philosophers sometimes make assertions like the following:

(i′)   The Epimenides proposition has no truth-value.
(ii′)   There is a proposition that can only be expressed in Greek, not
         English.
(iii′)   The proposition you express by saying, 'I'm not well' is distinct
         from the proposition I express by uttering the same sentence.

And yet, some of the philosophers who make such assertions also make
the assertion:

         There are no (such things as) propositions.

These facts suggest the need for a term for an implication to the effect
that there are Ss, whether or not accompanied by an assertion to the
effect that there are Ss, and even if accompanied by a denial that there are
Ss. We introduced informally (at the end of the fourth paragraph in this
book) 'T involves abstract entities' as short for 'T logically implies that
there exist entities of some sort that philosophers classify as abstract'.
Quine introduced 'T is ontologically committed to Ss' as short for 'T
logically implies that there exist Ss'. Several points concerning Quine's
usage call for comment.

First, the grammatical form of 'T is ontologically committed to Ss' may
suggest that if T is ontologically committed to Ss and Ss are Rs, then T is
ontologically committed to Rs. But this is a fallacious inference: if T logically implies that there are Ss, it does not follow that T logically implies
that there are Rs, even if Ss are in actual fact Rs. That only follows if
T logically implies that Ss are Rs (and then it follows whether or not Ss
are in actual fact Rs). When Ss are entities of a sort that philosophers
classify as abstract and T is ontologically committed to Ss, Quine will
speak loosely and say, 'T is ontologically committed to abstract entities',
even though T strictly speaking may not logically imply 'there are abstract
entities', and indeed usually will not if T is a scientific theory not involving the philosophical term 'abstract'. But this loose way of speaking is an
unofficial extension of the official sense of 'ontological commitment'.

Second, Quine holds that the only intelligible sense of truth is the
disquotational, and in consequence will use 'the existence of Ss follows
logically from T' interchangeably with '. . . follows logically from the
truth of T'. Likewise, Quine holds (as in Quine 1966c) that the only
intelligible sense of necessity is logical, and in consequence will use '. . . follows logically from . . .' interchangeably with '. . . follows necessarily from

. . .'. Accordingly, he will say that 'T is ontologically committed to Ss if and only if the existence of Ss follows necessarily from the truth of T, if and only if T cannot possibly be true unless there are Ss'. Given his distinctive and controversial views about the nature of truth and necessity, he means no more and no less by this than 'the existence of Ss follows logically from T, T implies that there are Ss', but his words may suggest something different to a reader who does not bear Quine's distinctive and controversial views in mind.

Third, there is a matter not of substance but of tone. While there may well be a need for some term for the notion in question, it is more doubtful whether there is a need for a term quite so solemn-sounding as the one Quine proposes. (Terms Quine uses for the same notion range from the somewhat less solemn 'posit' through 'reification' to the even more solemn 'hypostasis'.) The solemnity of the term has led to complaints that it tends to beg the question against those anti-nominalist views that (unlike Quine's anti-nominalism) hold the especial preoccupation of some philosophers specifically with ontological commitments to be inappropriate. We sympathize with such complaints, and we hope that our satiric intent in introducing such pretentious polysyllabic Hellenisms as 'syndynatontological' and 'hypocatastatic' has been recognized.

Ontological commitment has been a major theme in Quine's writings from the period of immediately after his collaboration with Goodman (Quine 1948) to quite late in his career (Quine 1981: §II). The main issue with which he has been concerned has been the search for a decisive test for ontological commitment, applicable even in the absence of assertion, or in the presence of denial. In Quine's usage, such a decisive test would be an **ontological criterion**: applied to a theory T involving a distinctive category of expressions, the S-expressions, an ontological criterion would tell us whether T correspondingly involves a distinctive category of entities, the S-entities. (If we wanted to be equally solemn about other matters, we might use the term **apheremenological criterion** for a decisive test for abstractness such as we failed to find in section I.A.1.)

Various aspects of the usage of S-expressions have been considered as candidate criteria. First, there is the **nominal** criterion, emphasizing the use of S-expressions as nouns, as in (i) above. Second, there is the **quantificational** criterion, emphasizing the role of assertions of the form:

there is an S that . . .

as in (ii) above. Third, there is the **identificational** criterion, emphasizing the role of assertions of the forms:

> . . . is the same S as . . .
> . . . is a different S from . . .
> there is one unique S that . . .
> there are several distinct Ss that . . .

as in (iii) above. All three candidate criteria are amply satisfied in the case of number-expressions.

As for the nominal criterion, though numerals were first used purely as adjectives, as in:

> Jane Austen wrote six novels

they are now used extensively as nouns, as in:

> Six is the number of novels Jane Austen wrote
> Six is a perfect number

(Kneale and Kneale (1963: chapter VI, §2) traces the transition to 'around the time of Plato'.) Today we have an elaborate system of numerals:

> one, two, three, . . .

and besides these, compound phrases formed using such expressions as:

> the number of . . .   the sum of . . .   the product of . . .

As for the quantificational criterion, we have both the particular and the universal:

> there is a number . . .   for every number . . .

as well as complicated iterations, alternations, and nestings:

> for every number m there is a number n . . .

These appear not only in the indicative in theorems, but also the interrogative in problems, and the imperative in algorithms:

> is there any number n . . .?   take any number n . . .!

And as for the identificational criterion, we have both equations and inequalities:

> . . . equals . . .   . . . is greater than . . .   . . . is less than . . .

as well as assertions of unicity and multiplicity:

> there exists a unique number . . .
> there exist several distinct numbers . . .

Now Quine has always disparaged the nominal criterion. By contrast, he has sometimes advanced slogans that sound like endorsements of the quantificational criterion or of the identificational criterion—'To be is to be the value of a bound variable', and 'No entity without identity'—which his critics have been quick to disparage. Must anyone who asserts that there is a strong chance that the whereabouts of the missing cash is the same as the whereabouts of the missing cashier, be deemed ontologically committed to such exotica as chances and whereaboutses? Actually, reservations about these criteria, too, appear in Quine's later works, and most likely have been present all along, even in earlier works where they are less conspicuous. His final view (as expressed in the later of the two works of his cited above) has been the pessimistic one that there is no criterion for drawing the line between the committed and the noncommittal, so far as theories expressed in natural languages like ordinary English are concerned:

. . . [T]here is no line to draw. Bodies are assumed, yes; they are the things, first and foremost. Beyond them there is a succession of dwindling analogies. Various expressions come to be used in ways more or less parallel to the use of terms for bodies, and it is felt that corresponding objects are more or less posited, pari passu; but there is no purpose in trying to mark an ontological limit to the dwindling parallelism.

My point is not that ordinary language is slipshod, slipshod though it be. We must recognize this grading off for what it is, and recognize that a fenced ontology is just not implicit in ordinary language.

The various candidate criteria point to various important 'ways of using terms for bodies', but none is claimed to be decisive.

Quine follows this pessimistic passage with a more optimistic one:

We can draw explicit ontological lines when desired. We can regiment our notation, admitting only general and singular terms, singular and plural predication, truth functions, and the machinery of relative clauses; or equivalently and more artificially, instead of plural predication and relative clauses we can admit quantification. Then it is that we can say that the objects assumed are the values of the variables, or of the pronouns. Various turns of phrase in ordinary language that seemed to invoke novel sorts of objects may disappear under such regimentation. At other points new ontic commitments may emerge. There is room for choice . . .

The positive part of Quine's view here is that while ontological commitment is a matter of 'more or less', of shades of grey, for natural language, it can be made a matter of 'all or none', of black and white, by suitable regimentation (whether or not one goes on to symbolization and formalization). For in a suitably regimented language, owing to the very limited range of grammatical and logical constructions and operations present, there will only be usages altogether like the usage of 'terms for bodies', and usages not at all like the usage of 'terms for bodies'. There will be no puzzling in-between cases.

There is, however, a negative aspect inconspicuously present even in the above positive-seeming passage. The negative part of Quine's view here is that the only suitable regimentation is what he sometimes calls **canonical** regimentation, where the only logical operators admitted are those of standard logic. No modal or other extended logics are admitted. Quine's most famous early discussion of ontological commitment (the earlier of the two papers of his cited above) began by saying that the question that interests him can be stated in three Anglo-Saxon monosyllables, 'What is there?' But he will listen to no answer involving those three fine old Anglo-Saxon monosyllables, the modal auxiliary verbs 'could, would, might'. Quine says that modal and other extensions of standard logic 'obstruct ontological comparison'. The ontological commitments of an uncanonically regimented theory are simply indeterminate, as much as or more so than those of an unregimented theory.

This view of Quine's is uniformly, if implicitly, rejected by all contemporary reconstructive nominalists, since they all make use of extended logics, and all claim thereby to be reducing ontological commitments. For that matter, very few if any anti-nominalists critical of reconstructive nominalist projects have defended Quine's views. Their criticisms of a strategy are often more of its non-logical than of its logical apparatus; when the criticism is of the logical apparatus, it is often more on ideological grounds of clarity than on ontological grounds; when the criticism is on ontological grounds, it is often more on grounds that there are determinate ontological commitments to abstracta implicit in the logic than on grounds that use of the logic makes ontological commitments indeterminate; and when the criticism is that ontological commitments are made indeterminate by some extended logic, this is often more a criticism of that one extended logic than a general claim like Quine's about any such logic.

To say that there is an implicit consensus against Quine's view is not, of course, to say that there is a consensus in favour of any one alternative view. Virtually all anti-nominalists and revolutionary nominalists hold, presumably on the basis of the features of the usage of number-expressions cited above, and perhaps also on the basis of the less philological consideration that there is an important body of theory explicitly called the 'theory of numbers', including an important body of results explicitly called 'existence theorems', that ordinary mathematics and science involve ontological commitments to numbers and the like. This hermeneutic nominalists deny. In denying it, in claiming that in this case the ostensible ontological commitments are spurious rather than genuine, they are obviously denying the sufficiency of the kinds of criteria cited above. Inversely, hermeneutic reductivists deny the necessity of those kinds of criteria. For they allege that ontological commitments are present, though latent, even where not manifested in any of the ways cited above. One side says its opponents suffer from ontological 'hallucinations' or 'mirages', and is frightened of imagined commitments that are quite unreal. The other side says its opponents are ontological 'myopics' or 'ostriches', unable or unwilling to see commitments that are right out in the open in front of them.

Fortunately there are at least some subsidiary points on which there is agreement at least among hermeneutic nominalists, whose views for the moment are the ones that concern us. For instance, Alonzo Church, in a paper generally supportive of Quine, Church (1950), gives some examples of philosophers who make assertions about propositions like (i′)–(iii′) above only to turn around and deny that there are any such things as propositions, and concludes that in the absence of further explanation such philosophers must be considered to have contradicted themselves. Most hermeneutic nominalists tacitly agree that this is so in the absence of further explanation, and for that reason engage in hermeneutic projects: their reconstruals of proposition-expressions, or more often of number-expressions, are intended to supply the further explanations required.

For another instance, William Alston, in a paper less sympathetic to Quine, Alston (1958), raises some questions precisely about nominalistic reconstruals. Consider the reinterpretation of:

(iv)   There are many virtues which he lacks.

as:

(v)   He might conceivably have been much more virtuous than he is.

The former appears, as the latter does not, to entail:

(vi)   There are such things as virtues.

Roughly and briefly, Alston's point is that the claim that the reinterpretation preserves meaning in itself gives as much reason to suspect that (v) really entails (vi) after all, as it does to doubt that (iv) ever really entailed (vi). This is because the relation of likeness of meaning or synonymy is a symmetric relation.

Again most hermeneutic nominalists tacitly agree (though whether they have fully taken in the significance of Alston's point is another matter). For they almost all, with one degree of explicitness or another and in one terminology or another, claim some asymmetric relation stronger than mere synonymy holds between an assertion seemingly about virtues, or more often numbers, and its nominalistic paraphrase. One fairly common way of putting the claim would be to say that (v) uncovers the **depth** form underlying the **superficial** form of (iv). This terminology derives from a distinction between 'deep' and 'surface' structure advocated at one time by Noam Chomsky; but it should be understood that the use of the terminology is not intended to indicate endorsement of any specific details of Chomsky's views at that or any other time. The distinction may be broadly Chomskyesque, but it is not intended to be narrowly Chomskyite.

With this understanding, and one further clarification, we will acquiesce in the 'superficial' vs. 'depth' metaphor ourselves. The further clarification is this: since presumably a sentence is one thing, and its form another, what it means to say that (v) 'uncovers the depth form underlying the surface form of' (iv) is not that (iv) and (v) are superficial and depth forms, respectively, but rather that they are sentences having the same depth form, which depth form is much less like the superficial form of (iv) than like that of (v).

b. *Reconstrual and 'Ontological Commitment'*

There is a fairly extensive range of usages in colloquial language that seem to imply or presuppose the existence of curious entities called 'chances':

(i)   There is a very strong chance that Professor Moriarty is responsible for this outrage.

(ii)   There is no strong chance that Professor Moriarty is responsible for this outrage, unless he had Colonel Moran as an accomplice.

(iii)   There is a quite strong chance that Professor Moriarty is responsible for this outrage, with Colonel Moran as an accomplice.

Here (i) gives an appearance of implying:

(iv)   There are such things as strong chances.

Much as:

(v)   There is a very sturdy chair that Professor Moriarty was reposing in this afternoon

implies:

(vi)   There are such things as sturdy chairs.

Of course, there is no important body of results explicitly called the 'chance theory', as there is a body of results called 'number theory'—or rather, there is something called the 'theory of chances', namely the theory of probability; but in that theory, the 'chance' or probability that P is just a number. Even anti-nominalists, therefore, are likely to be suspicious of the claim (iv). It may be wise to resign oneself to having to speak rather loosely in many everyday contexts, and nominalists and anti-nominalists alike have generally attached no very high priority to reconstruing colloquial talk about 'chances'. However, a nominalistic paraphrase of talk like (i)–(iii) does suggest itself:

(i*)   It's very likely that Professor Moriarty is responsible for this outrage.
(ii*)   It's not very likely that Professor Moriarty is responsible for this outrage, unless he had Colonel Moran as an accomplice.
(iii*) It's likely that Professor Moriarty is responsible for this outrage, with Colonel Moran as an accomplice.

The method * of paraphrase that carries (i)–(iii) to (i*)–(iii*) in general seems to preserve logical implications: (iii*) seems to follow from (i*) and (ii*) just as (iii) seems to follow from (i) and (ii). What happens when we apply * to (iv)?

The result would seem to be something like:

(iv*)   It's likely.

This, however, is not a proper English sentence, or at least not a complete one. And this may very well help convince one, if one wasn't convinced already, that the ontological thesis (iv), so far from being an implication of the everyday remark (i), isn't even a proper English sentence, but a bit of nonsense. And once this suggestion has occurred to us, we soon notice a difference between (i) and (v)—and not even a subtle one needing modern linguistics to detect it, but a gross one detectable already by traditional grammar. What follows 'that' in one case is a complete sentence, and in the other not.

Some of the nominalistic paraphrases considered earlier in this book can also help suggest that certain sentences that superficially appear to be in order, and to be consequences of ordinary beliefs, are in fact more or less nonsensical. For example, the assertion of actual existence:

(vii)   There are prime numbers greater than 200

appears to imply an assertion of possible existence:

(viii)   There either are or could have been prime numbers greater than 200

just as:

(ix)   There are human beings older than 200

implies:

(x)   There either are or could have been human beings older than 200.

Consider, however the modal paraphrase of Chapter II.B; call it †:

(vii†)   There could have been prime numeral tokens greater than 200.
(viii†)   There either could have been or could have could have been prime numeral tokens greater than 200.

Here the double 'could have' in (viii†) seems questionable English at best. And this may very well help convince one, if one wasn't convinced already, that applying the actual/possible distinction to numbers produces what is at best questionable sense.

But the issue we indicated in section o that we wished to take up in this section is a somewhat different one. Besides appearing, probably misleadingly, to imply (viii), the orthodox existence theorem (vii) also appears to imply the ontological thesis:

(xi)   There are numbers

which nominalists deny. The question we wished to take up in this section was whether (vii) only apparently implies (xi) or really does so. More precisely, the question we wished to take up was whether a nominalistic method of paraphrase can help to establish that the apparent implication is illusory and not real.

We will illustrate the issue by reference specifically to the purely modal reconstrual †, though the points we will be making are intended to apply quite generally. So let us apply the method of paraphrase † to (xi). We get:

(xi†)   There could have been numeral tokens.

Unlike (iv*) and (viii†), (xi†) does not look like nonsense. It looks like a logical implication of (vii†), and surely it is something that nominalists who accept the apparatus used in the paraphrase † want to affirm, and not to deny. But if the paraphrase (vii†) of (vii) implies the paraphrase (xi†) of (xi), does that fact not tend to reinforce rather than undermine the appearance that the original (vii) implies the original (xi)?

We have so far taken no note of claims to the effect that the paraphrase is a faithful analysis of the meaning of ordinary assertions, or of stronger claims to the effect that it uncovers depth forms underlying the superficial forms of ordinary assertions, in some Chomskyesque sense of 'depth' and 'superficial'. A Chomskyite linguist would surely ask just what evidence there is to support such a strong claim, but we indicated in section o that we would waive demands of evidence. We grant for the sake of argument any and all claims of this kind, however strong, a hermeneutic nominalist might wish to make. We cannot forbear, however, to point out the great historical irony in a nominalist's appeal to such claims. In the early literature of nominalism, anticipations of the thought that there may be something pertaining to the concrete somehow underlying assertions pertaining to the abstract are characteristic of the positions of anti-nominalists like Carnap and Dummett, with their 'confirmation-conditions', and 'verification-conditions', and the like. As for the nominalists, Goodman and Quine, they evinced a deep suspicion of appeal to any unobservable theoretical apparatus in language studies, a suspicion that aligned them with the 'structuralist' approach to linguistics of Bloomfield, and the 'behaviourist' approach of Skinner, and that was within a few years to lead them into a more than superficial conflict with Chomsky, and through him with the majority of scientific linguists.

Irony aside, there remains the question of relevance: let the claims about the relation between paraphrase and original be as strong as desired; still, so long as the paraphrase of a premiss appears to imply the paraphrase of a conclusion, appeal to the paraphrase will only serve to reinforce the appearance that the premiss implies the conclusion. Indeed, the stronger the relationship, the stronger the reinforcement. To be sure, considered in isolation, the alleged depth form (vii†) does not appear to imply (xi);

but the hermeneutic nominalist does not wish to consider these in isolation. On the contrary, the hermeneutic nominalist wishes to compare (vii†) with (vii) and claim a strong relation between the two. For the hermeneutic nominalist wishes to claim that (vii) is assertable because 'deep down it really only means' (vii†). Our point is that whatever evidence there may be for claiming that would seem to provide grounds for claiming something else, namely, that (xi) is assertable because 'deep down it really only means' (xi†). And the claim that (xi) is assertable is an anti-nominalist, not a nominalist claim.

We consider this point important enough that we will risk belabouring it by mentioning an analogy. Chomsky, at the time that he held his theory of deep and surface structures, held that deep structure is quite unlike surface structure in the case of sentences in the passive voice, such as:

(xii)   Chomsky's main critics are Bloomfield and Skinner; but
        Bloomfield has been refuted in Chomsky's *Syntactic Structures*, and
        Skinner has been refuted in Chomsky's review of *Verbal Behavior*.
(xiii) Chomsky's main critics have been refuted.

But Chomsky emphatically did not think that the alleged fact that the deep structures of these sentences are quite unlike their surface structures provided grounds for denying (xiii) and maintaining that Chomsky's main critics had after all *not* been refuted. Still less did he think that the alleged fact in question established the compatibility of affirmation of (xii) with denial of (xiii). But what goes for (xii) and (xiii) goes for (vii) and (xi) as well, we suggest.

Most of the points made so far (to the extent that they go beyond points made long ago by Alston in the work cited in article 2.a) are made with somewhat different emphases in Hodes (1990*b*). Hodes (as indicated in article B.2.c) himself makes strong hermeneutic claims to the effect that assertions not about numbers underlie ordinary assertions about numbers. He does not, however, conclude on that basis that the ordinary assertions are not 'ontologically committed' to numbers. Rather, he considers that when two levels of language are recognized, the notion of 'ontological commitment' has to be correspondingly divided into two notions he labels 'thin' and 'thick'.

Here 'thin ontological commitment' pertains to 'superficial' language, the only language one ever hears spoken or sees written, and the only language in which the early nominalists Goodman and Quine believed. Indeed 'thin ontological commitment' to numbers, for instance, is just 'ontological commitment' to numbers in Quine's sense, a willingness to assert, or to assert something implying, 'there are numbers'. Hodes himself is willing to undertake a 'thin ontological commitment' to numbers, and finds it easy to express this commitment in spoken or written 'surface' language simply by saying, like any anti-nominalist, that there are numbers. He will even say what some anti-nominalists would not—Carnap and Dummett probably would, Quine probably wouldn't—that it is a trivial truism that there are numbers. Thus early in his paper he writes:

The answers to questions like 'Are there numbers?' and 'Do sets exist?' are, trivially, 'Yes'. To not see these answers as trivialities bespeaks a misunderstanding of mathematical discourse.

To refuse to undertake 'thin ontological commitment' to numbers would be incompatible with current mathematics and science and common sense, according to Hodes.

By contrast, 'thick ontological commitment' pertains to 'depth' language, an unobservable theoretical posit of technical linguistic science. To undertake a 'thick ontological commitment' to numbers, for instance, would be to assert something whose 'depth' form is 'there are numbers', or to assert something implying such a thing. Hodes takes some of Quine's characterizations of 'ontological commitment', namely, formulations in terms of what has to exist in order for something to be true, to be characterizations of the 'thick' rather than the 'thin' kind. This is not so if (as discussed in article 2.a) the key modal construction 'has to . . . in order that . . .' and the alethic notion 'is true' are both understood as Quine understands them; but it may be so if one or the other is understood as Hodes understands it. Hodes is unwilling to undertake 'thick ontological commitment' to numbers, and holds that his refusal to do so is compatible with current mathematics and science and common sense.

It is not so easy, however, to express this refusal in 'surface' language, spoken or written. When Hodes tries to do so, in the continuation of the passage just quoted, what he is driving at can be understood well enough, to a first approximation, given a background explanation of the 'thin' vs. 'thick' distinction, such as we have just attempted to supply. Considered in isolation, however, it is hardly more self-explanatory than the language of those theorists (discussed in article A.2.c) who would make a distinction between first-rate 'existence' and second-rate 'being':

But to go on and say that there is a realm of mathematical objects is to engage in obscurantist hyperbole. Mathematical objects are second-rate; they are not among the 'furniture of the universe'.

Though there is no question of Hodes himself being confused, his non-self-explanatory language may well be confusing.

A far less confusing usage would be to use 'ontological commitment' only in its original, Quinine sense of existential implication, and introduce a new term, say **bathyontological commitment**, for what Hodes calls 'thick' ontological commitment. Along with it might be introduced a new term, say **infranominalism**, for the unwillingness to undertake such commitments to abstracta. The neologistic character of the terminology seems appropriate given the novel character of the issue, which has little directly to do with the original issues of nominalism, except in so far as in some vague and inchoate way early anti-nominalism anticipated infranominalism, while early nominalism was suspicious of the whole distinction between the 'bathy-' and 'phanero-' levels of analysis. The least confusing usage of all, however, would be to retire 'ontological commitment' and all its cognates and derivatives. Though the introduction of the phrase was motivated by a desire to increase clarity, experience shows that use of the phrase has tended to have the opposite effect.

That said, we will leave the issue of infranominalism to the professional linguists, and return to the issue we have been at pains to distinguish from it, that of nominalism. (In so doing, we will revert to our policy, announced in article I.A.o.a and maintained until section o above, of dropping qualifications like 'apparently' or 'superficially', and systematically ignoring the hermeneutic position.)

3. ENVOI: RECONSTRUAL WITHOUT NOMINALISM

Inimicus Plato, sed magis inimica falsitas.
—Tarski

What does the product of the work of reconstructive nominalists, the array of reconstruals or reconstructions surveyed in this book, look like from an anti-nominalist viewpoint? Since anti-nominalists reject all hermeneutic and revolutionary claims, from their viewpoint the various reconstruals or reconstructions are all distinct from and inferior to current theories. What is accomplished by producing a series of such distinct and inferior theories? No advancement of science proper, certainly; but perhaps a contribution to the philosophical understanding of the character of science.

Putnam, in his discussion of 'equivalent descriptions' (mentioned in article B.2.d), already suggests some ways in which the availability of equivalent descriptions in general, and nominalistic alternatives in particular, may enlarge philosophical understanding and contribute to naturalized epistemology and cognitive studies. Let us elaborate on some of his general suggestions, without pretending to remain faithful to his particular emphases.

First, there is a contribution to the solution of the problem of how current abstractly formulated scientific beliefs and assertions could have been arrived at. Note that this is a problem for nominalists and antinominalist alike: how the beliefs and assertions were arrived at must be explained whether one holds them to be *justified* beliefs and *warranted* assertions or not. The various nominalistic paraphrases suggest various possible routes: one might start with modally formulated beliefs and assertions, or with beliefs and assertions formulated in terms of proportionality relationships as in synthetic geometry, and then arrive at abstractly and numerically formulated beliefs by a certain linguistic transformation, namely, the nominalistic paraphrase run in reverse. When one recalls (as we did briefly in sections II.A.o and II.B.o) that mathematics in significant part actually was originally formulated in terms of what is proportional to what or in terms of what it is possible to construct, rather than in terms of what numerical ratios or other abstracta exist, one can say that the nominalistic paraphrases point to not just a possible route towards what are now the standard formulations, but also what was in part the actual route.

Second, there is a contribution to answering the question *why* the linguistic transformations alluded to were made. Note again that this is a question that arises for nominalists and anti-nominalist alike: why the transformations were made must be explained whether one regards that development as progress or as error. And though it may seem a back-handed kind of compliment to say so, by their very awkwardness and inconvenience the nominalistic strategies make a real contribution to explaining why the linguistic transformations alluded to were practically unavoidable if science was to develop: they demonstrate as nothing else does just how much more convenient and perspicuous a numerical or otherwise abstract formulation can be. Of course, the nominalistic strategies do nothing by themselves to explain *why* numerical and otherwise abstract formulations are so much less awkward and so much more perspicuous; but they at least serve to call attention as nothing else does to this interesting psychological fact about cognition.

Third, there is a contribution to naturalized epistemology of a more fundamental kind, connected with one of the most important general issues for the philosophical understanding of the character of science. To explain the nature of the contribution will require some preliminary discussion of the general issue in question.

We are intelligences embodied in a physical universe of a certain kind, organisms of a certain biological species, bearers of a certain historical and social and cultural tradition, individuals each with psychological peculiarities of his or her own, and subjects of accidents happy or unhappy and luck good or bad. One of the most important general issues for the philosophical understanding of the character of science is just this: to what extent does the way we are, rather than the way the world of numerical and material and living entities is, shape our mathematical and physical and biological theories of the world? Discussion of this question has long been dominated by extreme views that radically minimize or virtually deny either the contribution from us or the contribution from facts not of our making about the world.

Thus the early modern pioneers in the development of the sciences tended (as we mentioned in passing in article I.A.2.a) to regard their scientific theories of the world as directly corresponding to reality, which for them as theists or deists meant corresponding to the Creator's design for the world. The standard history of mathematics, Kline (1972), collects many expressions of this attitude by Kepler, Galileo, Descartes, and other worthies of the period; but the best description of the attitude in the collection is perhaps that given by one who did not share it, William James, in the early pages of a work devoted to arguing against it, *Pragmatism*:

When the first mathematical, logical, and natural uniformities, the first laws, were discovered, men were so carried away by the clearness, beauty and simplification that resulted, that they believed themselves to have deciphered authentically the eternal thoughts of the Almighty. . . . He also thought in conic sections, squares and roots and ratios, and geometrized like Euclid. He made Kepler's laws for the planets to follow; he made velocity increase proportionally to the time in falling bodies [Galileo's law]; he made the law of sines [of which Descartes was a co-discoverer] for light to obey when refracted; . . . and when we rediscover any one of these his wondrous institutions, we seize his mind in its very literal intention.

By now, however, it has long since become a truism that our biological and physical and mathematical theories of life and matter and number are to a significant degree shaped by our character, and in particular by our history and our society and our culture.

What today is in danger of being overlooked or outright denied is not the role of history or society or culture in shaping scientific theory, but rather the role of mathematical and physical and biological facts not of our making. For the most fashionable figures in the history and sociology and anthropology of science deny not only that there is a ready-made theory of the world, but even there is any ready-made world. They maintain not just that theories about life and matter and number are constructs of human history and society and culture, but that number and matter and life themselves are such constructs. In so maintaining, they already contradict accepted scientific estimates that life and matter are billions of years older than the human species, and contravene the accepted mathematical treatment of number as timeless. But their opposition to science goes further. For they are quick to go on to conclude that mathematical and physical and biological facts, being created by us when we create mathematical and physical and biological theories, cannot impose any prior constraint on how we go about shaping those theories, leaving only constraints from our side—assumed to be social and political and economic—rather than the world's side.

The canonical expression of this attitude has been given by the trendiest sociologist of science of them all, Bruno Latour, in his 'Third Rule of Method', reiterated several times in the course of Latour (1985) and other works. The rule reads as follows:

Since the settlement of a controversy is the Cause of Nature's representation, not the consequence, we can never use the outcome—Nature—to explain how and why a controversy has been settled.

(For quotations and discussion of many similar formulations by other modish thinkers, see Laudan (1992) and Gross and Levitt (1994). It is because of the vogue for views that dismiss the reasons scientists give for coming down on one side of a controversial question rather than another as mere rationalizations for social power that contemporary history of science produces so little of use in connection with issues like the one we were considering in article 1.b.)

It is unlikely that any reader of a book like this one will be a sympathizer with views like Latour's, so it is probably superfluous for us to argue at any length against such views (which one of us has already addressed in Burgess (1993: §3)). However, for the record, let us note that the quoted argument is an impudent sophism, quite worthy of Euthydemus, a fallacy of equivocation conflating 'Nature' with 'Nature's representation'. Moreover, its conclusion gets the relationship between science and power almost exactly backwards: the answers science gives to the questions it considers are only constrained to a very slight degree by political and economic power, but are constrained to a very high degree by regularities in the world no political or economic power can change; and it is for that very reason that the answers science delivers are applicable to the world to such a high degree, and capable for good or ill of conferring so much further power on whoever already has enough of it to be able to influence what questions science considers and what applications are made of its answers.

Serious thought about the character of science can only begin when both the older faith that science can provide a God's eye view of the universe and the currently chic opinion that science is 'just another narrative that reinscribes relations of social dominance' are alike dismissed, and it is acknowledged that science is shaped both by ourselves and by facts not of our making. The question then arises, how much is it shaped by the one, and how much by the other? And as for factors coming from our side rather than the world's side, how much is due to one kind of factor rather than another?

Quine—to quote him one last time—has noted a difficulty with this kind of question. Towards the end of Quine (1950) he writes of a closely similar question:

> The fundamental-seeming philosophical question, How much of our science is merely contributed by language and how much is a genuine reflection of reality? is perhaps a spurious question . . . Certainly we are in a predicament if we try to answer the question; for to answer the question we must talk about the world as well as about language, and to talk about the world we must already impose upon the world some conceptual scheme peculiar to our own special language.

Quine—to express partial agreement and partial disagreement with him for one last time—has noted a genuine difficulty here, but overlooked a partial solution. Quine is quite right in so far as he suggests that what we contribute cannot be isolated by producing a theory of the world uncontaminated by any contribution from us, and comparing our actual scientific theories to it. Yet one possible way one might hope to gain insight into what and how much we contribute remains, and Quine seems wrong in so far as he seems to neglect this possibility.

For one could hope to obtain such insight by producing a theory of the world that, though it no more than any other theory directly 'reflects' reality without the imposition of any 'conceptual scheme', does impose a different 'conceptual scheme' from that imposed by our actual scientific theories. Using the theory might be quite inconvenient or even unfeasible for us; but provided that it would in principle be possible for intelligences unlike us and carrying different biological and social and psychological baggage from ours, comparison of the theory with our actual scientific theories would help give a sense of what and how much our character has contributed to shaping the latter. Devising alternatives distinct from and inferior by our standards to our actual theories, but in principle possible to use in their place, is a way of imaging what the science of alien intelligences might be like, and as such a way of advancing the philosophical understanding of the character of science.

It is just such an advance, we want to suggest, that is accomplished by the various reconstructive nominalistic strategies surveyed in this book. And this is an accomplishment that can and ought to be recognized even by anti-nominalists. Indeed, the anti-nominalist is better placed than the hermeneutic or revolutionary partisans of any one particular strategy to appreciate the value of the variety of reconstruals and reconstructions produced by different strategies. For at most one strategy could produce a faithful exegesis of current mathematical and scientific theories; at most one could produce a scientifically optimal replacement for them; but many can contribute to enlarging our understanding of the character of our science by showing what the science of other intelligent creatures might be like.

The advance accomplished by the various reconstructive strategies in the direction that most interests nominalists may lead to advances in other directions as well. Within the field of philosophy of mathematics, what has been accomplished in connection with the issue of nominalism provides a model for what one could hope to accomplish in connection with the issue of constructivism, though here different tools will be needed.

Constructivist heretics tend to be even more intolerant of nominalist heresy than are the orthodox. Constructivists tend to claim that the philosophical issues connected with constructivism are more profound than those connected with nominalism, and that the technical work connected with constructivism is deeper. While the former claim is debatable, the latter is indisputable. Yet all the indisputably deep technical work associated with constructivism pertains only to the reconstrual or reconstruction of purely mathematical theories: the reconstrual or reconstruction of mixed mathematico-physical theories remains a task to be undertaken. The work of nominalists, if it accomplishes nothing else, at least serves to call attention to this omission. It is all very well to say that producing a constructivist physics would be a philosophically more profound achievement than producing a nominalist physics; but the production of a constructivist physics is a hope for the future, while the production of a nominalist physics is an actual accomplishment, at least to the extent surveyed in this book.

How much does the way we are, rather than the way the world is, help shape our theories of the world: how might our theories of the world have been different if we had been different, but the world the same? This, we have said, is one of the most important general issues for any serious thought about the character of science. (Perhaps we should add one more time just for emphasis that recognition that our theories might have been different if we had been different and the world the same in itself at most suggests that our theories are not *uniquely right*, and does not establish that our theories are not *all right*.) The various strategies for nominalistic reconstrual of mathematics surveyed in this book together provide a partial answer, but of course only a partial answer, pertaining only to one aspect of the question.

It is well, therefore, to hope that those who find other aspects of the question deeper and more interesting will undertake to imitate and if possible surpass what has been thus achieved by reconstructive nominalists in connection with the aspect of the question that (for reasons cogent or fallacious) has most interested them. But though it is well to hope thus for more, it is also well to be thankful for what has already been provided.

REFERENCES

Many of the references consulted have been anthologized and reprinted, some many times, and where this is so we have generally listed both the original publication and the reprint most likely to be available to readers. In the body of this book, citations of specific passages in the references are generally given by internal divisions in the works in question, as in 'Goodman (1956: §3, objection (iv))', since these remain the same from one printing to another, while pagination varies.

ADAMS, MARILYN (1982), 'Universals in the Early Fourteenth Century', in N. Kretzmann, A. Kenny, and K. Pinborg (eds.), *The Cambridge History of Later Medieval Philosophy* (Cambridge), 411–39.

ALSTON, WILLIAM (1958), 'Ontological Commitments', *Philosophical Studies*, 9: 8–17; repr. in Benacerraf and Putnam (1964).

ARMSTRONG, DAVID (1978), *Universals and Scientific Realism* (2 vols.; Cambridge).

—— (1991), 'Classes Are States of Affairs', *Mind*, 100: 189–200.

BACON, JOHN (1985), 'The Completeness of Predicate Functor Logic', *Journal of Symbolic Logic*, 50: 903–26.

BENACERRAF, PAUL (1965), 'What Numbers Could Not Be', *Philosophical Review*, 74: 47–73; repr. in Benacerraf and Putnam (1983).

—— (1973), 'Mathematical Truth', *Journal of Philosophy*, 70: 661–80; repr. in Benacerraf and Putnam (1983).

—— and PUTNAM, H. (1964) (eds.), *Philosophy of Mathematics: Selected Readings*, 1st edn. (Englewood Cliffs, NJ).

—— —— (1983) (eds.), *Philosophy of Mathematics: Selected Readings*, 2nd edn. (Cambridge).

BERNAYS, PAUL (1935), 'Sur le platonisme dans les mathématiques', *L'Enseignement Mathématique*, 34: 52–69; trans. C. D. Parsons, 'On Platonism in Mathematics', in Benacerraf and Putnam (1964, 1983).

BIGELOW, JOHN (1988), *The Reality of Numbers: A Physicalist's Philosophy of Mathematics* (Oxford).

—— (1990), 'Sets Are Universals', in Irvine 1990: 291–306.

BONEVAC, DANIEL (1982), *Reduction in the Abstract Sciences* (Indianapolis).

—— (1983), 'Freedom and Truth in Mathematics', *Erkenntnis*, 20: 93–102.

—— (1984), 'Systems of Substitutional Semantics', *Philosophy of Science*, 51: 631–56.

—— (1985), 'Quantity and Quantification', *Noûs*, 19: 229–47.

BOOLOS, GEORGE (1984), 'To Be is To Be the Value of a Variable (or To Be Some Values of Some Variables)', *Journal of Philosophy*, 81: 430–9.

Bostock, David (1974–9), *Logic and Arithmetic* (2 vols.; Oxford).

Bull, R. A., and Segerberg, K. (1984), 'Basic Modal Logic', in Gabbay and Guenthner 1984: 1–88.

Burgess, John P. (1983), 'Why I Am Not a Nominalist', *Notre Dame Journal of Formal Logic*, 24: 93–105.

—— (1984*a*), 'Basic Tense Logic', in Gabbay and Guenthner 1984: 89–134.

—— (1984*b*), 'Synthetic Mechanics', *Journal of Philosophical Logic*, 13: 379–95.

—— (1989), 'Sets and Point-Sets', in A. Fine and J. Lepin (eds.), *PSA 1988: Proceedings of the 1988 Biennial Meeting of the Philosophy of Science Association*, ii. 456–63.

—— (1990*a*), 'Synthetic Mechanics Revisited', *Journal of Philosophical Logic*, 20: 121–30.

—— (1990*b*), 'Epistemology and Nominalism', in Irvine 1990: 1–15.

—— (1991), 'Synthetic Physics and Nominalist Realism', in C. W. Savage and P. Ehrlich (eds.), *Philosophical and Foundational Issues in Measurement Theory* (Hillsdale, Mich.), 119–38.

—— (1992), 'Proofs About Proofs: A Defense of Classical Logic, I', in M. Detlefsen (ed.), *Proof, Logic and Formalization* (London), 8–23.

—— (1993), 'How Foundational Work in Mathematics Can be Relevant to Philosophy of Science', in D. Hull, M. Forbes, and K. Okruhlik (eds.), *PSA 1992: Proceedings of the 1992 Biennial Meeting of the Philosophy of Science Association*, ii. 433–41.

—— (1995), 'Non-Classical Logic and Ontological Non-Commitment: Avoiding Abstract Objects through Modal Operators', in D. Prawitz, B. Skyrms, and D. Westerståhl (eds.), *Logic, Methodology and Philosophy of Science IX* (Amsterdam), 287–305.

Carnap, Rudolf (1950), 'Empiricism, Semantics, and Ontology', *Revue internationale de philosophie*, 4: 20–40; repr. in Benacerraf and Putnam (1964, 1983).

Chihara, Charles (1973), *Ontology and the Vicious Circle Principle* (Ithaca, NY).

—— (1984), 'A Simple Type Theory without Platonic Domains', *Journal of Philosophical Logic*, 13: 249–83.

—— (1990), *Constructibility and Mathematical Existence* (Oxford).

Church, Alonzo (1950), 'Ontological Commitment', *Journal of Philosophy*, 55: 1008–14.

Dummett, Michael (1956), 'Nominalism', *Philosophical Review*, 65: 491–505; repr. in the author's *Truth and Other Enigmas* (Cambridge, Mass., 1978).

Earman, John (1970), 'Who's Afraid of Absolute Space?' *Australian Journal of Philosophy*, 48: 287–319.

Field, Hartry H. (1972), 'Tarski's Theory of Truth', *Journal of Philosophy*, 69: 347–75.

—— (1980), *Science without Numbers: A Defence of Nominalism* (Princeton).

—— (1982), 'Realism and Anti-Realism about Mathematics', *Philosophical Topics*, 13: 45–69.

—— (1984*a*), 'Is Mathematical Knowledge Just Logical Knowledge?' *Philosophical Review*, 93: 509–52.

—— (1984*b*), Review of Wright (1983), *Canadian Journal of Philosophy*, 14: 637–62.

—— (1985*a*), 'On Conservativeness and Incompleteness', *Journal of Philosophy*, 81: 239–60.

—— (1985*b*), 'Can We Dispense with Space-Time?' in P. Asquith and P. Kitcher (eds.), *PSA 1988: Proceedings of the 1988 Biennial Meeting of the Philosophy of Science Association*, ii. 30–90.

—— (1986), 'The Deflationary Conception of Truth', in G. McDonald and C. Wright (eds.), *Fact, Science and Value: Essays on A. J. Ayer's Language, Truth and Logic* (Oxford).

—— (1988), 'Realism, Mathematics and Modality', *Philosophical Topics*, 19: 57–107.

—— (1989), *Realism, Mathematics and Modality* (Oxford).

—— (1990), 'Mathematics without Truth (A Reply to Maddy)', *Pacific Philosophical Quarterly*, 71: 206–22.

—— (1991), 'Metalogic and Modality', *Philosophical Studies*, 62: 1–22.

—— (1992), 'A Nominalistic Proof of the Conservativeness of Set Theory', *Journal of Philosophical Logic*, 21: 11–23.

GABBAY, D., and GUENTHNER, F. (1984) (eds.), *Extensions of Classical Logic*, being *Handbook of Philosophical Logic*, vol. ii (Dordrecht).

GARDNER, MARTIN (1989), *Science: Good, Bad, and Bogus*, paperback edn. (Buffalo, Colo.).

GARSON, JAMES W. (1984), 'Quantification in Modal Logic', in Gabbay and Guenthner 1984: 249–308.

GETTIER, EDMUND (1963), 'Is Justified True Belief Knowledge?' *Analysis*, 23: 121–3.

GÖDEL, KURT (1947), 'What Is Cantor's Continuum Problem?' *American Mathematical Monthly*, 54: 515–25; revised version repr. in Benacerraf and Putnam (1964, 1983).

GOLDMAN, ALVIN (1967), 'A Causal Theory of Knowing', *Journal of Philosophy*, 64: 357–72.

GOODMAN, NELSON (1956), 'A World of Individuals', in *The Problem of Universals: A Symposium* (Notre Dame, Ind.); repr. in Benacerraf and Putnam (1964).

—— and LEONARD, H. (1940), 'The Calculus of Individuals and Its Uses', *Journal of Symbolic Logic*, 5: 45–55.

—— and QUINE, W. V. (1947), 'Steps toward a Constructive Nominalism', *Journal of Symbolic Logic*, 12: 97–122; repr. in the first author's *Problems and Projects* (Indianapolis, 1972).

GOTTLIEB, DALE (1980), *Ontological Economy: Substitutional Quantification and Mathematics* (Oxford).

GROSS, P., and LEVITT, N. (1994), *Higher Superstition: The Academic Left and Its Quarrels with Science* (Baltimore).

GROVER, DOROTHY (1992), *A Prosentential Theory of Truth* (Princeton).

HAHN, E., and SCHILPP, P. A. (1986) (eds.), *The Philosophy of W. V. Quine*, Library of Living Philosophers (La Salle, Ill.).

HALE, BOB (1987), *Abstract Objects* (Oxford).

—— (1990), 'Nominalism', in Irvine 1990: 121–44.

HALMOS, PAUL (1960), *Naive Set Theory* (Princeton).

HART, W. D. (1977), review of Steiner (1975), *Journal of Philosophy*, 74: 118–29.

HELLMAN, GEOFFREY (1989), *Mathematics without Numbers* (Oxford).

—— (1990), 'Modal-Structural Mathematics', in Irvine 1990: 307–30.

HILBERT, DAVID (1900), *Grundlagen der Geometrie* (Leipzig); trans. E. J. Townsend *The Foundations of Geometry* (Chicago, 1902).

HODES, HAROLD (1984a), 'On Modal Logics Which Enrich First Order S5', *Journal of Philosophical Logic*, 13: 123–49.

—— (1984b), 'Logicism and the Ontological Commitments of Arithmetic', *Journal of Philosophy*, 123–49.

—— (1990a), 'Where do Natural Numbers Come from?' *Synthese*, 84: 347–407.

—— (1990b), 'Ontological Commitment: Thick and Thin', in G. Boolos (ed.), *Essays in Honor of Hilary Putnam* (Cambridge), 235–60.

—— (1991), 'Where do Sets Come from?' *Journal of Symbolic Logic*, 56: 150–75.

HORWICH, PAUL (1990), *Truth* (Oxford).

IRVINE, ANDREW (1990) (ed.), *Physicalism in Mathematics* (Dordrecht).

JUBIEN, MICHAEL (1977), 'Ontology and Mathematical Truth', *Noûs*, 11: 133–50.

KEISLER, H. JEROME (1976), *Elementary Calculus: An Infinitesimal Approach* (Boston).

KITCHER, PHILIP (1978), 'The Plight of the Platonist', *Noûs*, 12: 119–36.

—— (1984), *The Nature of Mathematical Knowledge* (Oxford).

KLINE, MORRIS (1972), *Mathematical Thought from Ancient to Modern Times* (Oxford).

KNEALE, W., and KNEALE, M. (1963), *The Development of Logic* (Oxford).

KNUTH, DONALD W. (1974), *Surreal Numbers: How Two Ex-Students Turned on to Pure Mathematics and Found Total Happiness* (Reading).

KRIPKE, SAUL (1972), 'Naming and Necessity', in D. Davidson and G. Harman (eds.), *Semantics of Natural Language* (Dordrecht).

—— (1976), 'Is There a Problem about Substitutional Quantification?' in G. Evans and J. McDowell (eds.), *Essays in Semantics* (Oxford).

—— (1982), *Wittgenstein on Rules and Private Language* (Cambridge, Mass.).

LATOUR, BRUNO (1985), *Science in Action: How to Follow Scientists and Engineers through Society* (Cambridge, Mass.).

LAUDAN, RACHEL (1992), 'The "New" History of Science: Implications for Philosophy of Science', in D. Hull, M. Forbes, and K. Okruhlik (eds.), *PSA 1992: Proceedings of the 1992 Biennial Meeting of the Philosophy of Science Association*, ii. 476–81.

LEWIS, DAVID K. (1984), 'Putnam's Paradox', *Australian Journal of Philosophy*, 62: 221–36.

—— (1986), *On the Plurality of Worlds* (Oxford).

—— (1990), 'Noneism or Allism?' *Mind*, 99: 23–31.

—— (1991), *Parts of Classes* (Oxford).

—— (1993), 'Mathematics is Megethology', *Philosophia Mathematica*, 3rd ser. 1: 3–23.

MADDY, PENELOPE (1984*a*), 'Mathematical Epistemology: What Is the Question?' *The Monist*, 67: 46–55.

—— (1984*b*), 'How the Causal Theorist Follows a Rule', *Midwest Studies in Philosophy*, 9: 457–77.

—— (1988), 'Believing the Axioms' (in 2 parts), *Journal of Symbolic Logic*, 53: 481–511, 736–64.

—— (1990*a*), *Realism in Mathematics* (Oxford).

—— (1990*b*), 'Physicalist Platonism', in Irvine 1990: 259–89.

—— (1990*c*), 'Mathematics and Oliver Twist', *Pacific Philosophical Quarterly*, 71: 189–205.

NOETHER, E., and CAVAILLÈS, J. (1937) (eds.), *Briefwechsel Cantor-Dedekind* (Paris).

NORMORE, CALVIN (1987), 'The Tradition of Medieval Nominalism', in J. Wippel (ed.), *Studies in Medieval Philosophy* (Washington).

PARSONS, CHARLES (1986), 'Quine on the Philosophy of Mathematics', in Hahn and Schilpp 1986: 369–95.

PRIOR, ARTHUR (1969), *Past, Present and Future* (Oxford).

PUTNAM, HILARY (1967), 'Mathematics without Foundations', *Journal of Philosophy*, 64: 5–22; repr. in Benacerraf and Putnam (1983).

—— (1971), *Philosophy of Logic* (New York).

—— (1980), 'Models and Reality', *Journal of Symbolic Logic*, 45: 464–82; repr. in Benacerraf and Putnam (1983).

QUINE, WILLARD VAN ORMAN (1948), 'On What There Is', *Review of Metaphysics*, 2: 21–38; repr. in Benacerraf and Putnam (1964).

—— (1950), 'Identity, Ostension, Hypostasis', *Journal of Philosophy*, 47: 621–33; repr. in Quine (1953).

—— (1951*a*), 'On Carnap's Views on Ontology', *Philosophical Studies*, 2: 65–72; repr. in Quine (1966*a*).

—— (1951*b*), 'Two Dogmas of Empiricism', *Philosophical Review*, 60; repr. in Quine (1953) and in Benacerraf and Putnam (1964).

—— (1953), *From a Logical Point of View* (New York).

—— (1960*a*), *Word and Object* (Cambridge, Mass.).

—— (1960*b*), 'Variables Explained Away', *Proceedings of the American Philosophical Society*, 104: 343–7; repr. in the author's *Selected Logic Papers* (New York).

—— (1966*a*), *The Ways of Paradox* (New York).

—— (1966*b*), 'Posits and Reality', in Quine (1966*a*), 233–41.

—— (1966*c*), 'Necessary Truth', in Quine (1966*a*).

References

QUINE, WILLARD VAN ORMAN (1969), 'Epistemology Naturalized', in the author's *Ontological Relativity and Other Essays* (New York).

—— (1974), *The Roots of Reference* (La Salle, Ill.).

—— (1981), 'Things and Their Place in Theories', in the author's *Theories and Things* (Cambridge, Mass.).

RESNIK, MICHAEL (1981), 'Mathematics as a Science of Patterns: Ontology and Reference', *Noûs*, 15: 529–50.

—— (1982), 'Mathematics as a Science of Patterns: Epistemology', *Noûs*, 16: 95–105.

—— (1985a), 'How Nominalist is Hartry Field's Nominalism?' *Philosophical Studies*, 47: 163–81.

—— (1985b), 'Ontology and Logic: Remarks on Hartry Field's Anti-Platonist Philosophy of Mathematics', *History and Philosophy of Logic*, 6: 191–209.

—— (1988), 'Second-Order Logic Still Wild', *Journal of Philosophy*, 85: 75–87.

—— (1990), 'Beliefs about Mathematical Objects', in Irvine 1990: 41–72.

ROBB, ALFRED (1914), *A Theory of Time and Space* (Cambridge).

ROSEN, GIDEON (1990), 'Modal Fictionalism', *Mind*, 99: 327–54.

—— (1992), 'Remarks on Modern Nominalism', Ph.D. dissertation (Princeton).

—— (1993a), 'The Refutation of Nominalism (?)', *Philosophical Topics*, 21: 149–86.

—— (1993b), 'A Problem for Fictionalism about Possible Worlds', *Analysis*, 53: 71–81.

—— (1994), 'What is Constructive Empiricism?' *Philosophical Studies*, 74: 143–78.

—— (1995), 'Armstrong on Classes as States of Affairs', *Australasian Journal of Philosophy*, 73: 613–25.

SHAPIRO, STEWART (1983a), 'Mathematics and Reality', *Philosophy of Science*, 50: 523–48.

—— (1983b), 'Conservativeness and Incompleteness', *Journal of Philosophy*, 81: 521–31.

SHEARD, MICHAEL (1994), 'A Guide to Truth Predicates in the Modern Era', *Journal of Symbolic Logic*, 59: 1032–54.

SIMONS, PETER (1987), *Parts: A Study in Ontology* (Oxford).

SKLAR, LAWRENCE (1974), *Space, Time, and Spacetime* (Berkeley and Los Angeles).

SOAMES, SCOTT (1985), 'Semantics and Psychology', in J. Katz (ed.), *The Philosophy of Linguistics* (Oxford).

STEINER, MARK (1975), *Mathematical Knowledge* (Ithaca, NY).

TAIT, WILLIAM (1986), 'Truth and Proof: The Platonism of Mathematics', *Synthese*, 69: 341–70.

TARSKI, ALFRED (1959), 'What Is Elementary Geometry?' in L. Henkin, P. Suppes, and A. Tarski (eds.), *The Axiomatic Method with Special Reference to Geometry and Physics* (Amsterdam); repr. in J. Hintikka (ed.), *The Philosophy of Mathematics* (Oxford, 1969).

—— MOSTOWSKI, A., and ROBINSON, R. M. (1953), *Undecidable Theories* (Amsterdam).

—— and SZCZERBA, L. (1965), 'Metamathematical Properties of Some Affine Geometries', in J. Bar-Hillel (ed.), *Logic, Methodology, and Philosophy of Science* (Amsterdam), 166–78.

URQUHART, ALASDAIR (1990), 'The Logic of Physical Theory', in Irvine 1990: 145–54.

VAN BENTHEM, JOHAN (1977), 'Tense Logic and Standard Logic', *Logique et Analyse*, 80: 41–83.

VAN FRAASSEN, BAS (1980), *The Scientific Image* (Oxford).

WRIGHT, CRISPIN (1983), *Frege's Conception of Numbers as Objects* (Aberdeen).

—— (1990), 'Field and Fregean Platonism', in Irvine 1990: 73–93.

ZERMELO, ERNST (1930), 'Über Grenzzahlen und Mengenbereiche', *Fundamenta Mathematicæ*, 16: 29–47.

SECTION INDEX

The first occurrence of any term used in the body of this book in a special or technical sense can be located with the aid of the following index. There it will be found printed in **boldface** type, and accompanied by an explanation of the sense in which it is being used. The index also gives the places where the works of the various authors cited are quoted or discussed. Locations are given by part, chapter, section, and article.